Selected Titles in This Series

646 **Russell Johnson and Mahesh Nerurkar,** Controllability, stabilization, and the regulator problem for random differential systems, 1998

645 **Peter W. Bates, Kening Lu, and Chongchun Zeng,** Existence and persistence of invariant manifolds for semiflows in Banach space, 1998

644 **Michael David Weiner,** Bosonic construction of vertex operator para-algebras from symplectic affine Kac-Moody algebras, 1998

643 **Józef Dodziuk and Jay Jorgenson,** Spectral asymptotics on degenerating hyperbolic 3-manifolds, 1998

642 **Chu Wenchang,** Basic almost-poised hypergeometric series, 1998

641 **W. Bulla, F. Gesztesy, H. Holden, and G. Teschl,** Algebro-geometric quasi-periodic finite-gap solutions of the Toda and Kac-van Moerbeke hierarchies, 1998

640 **Xingde Dai and David R. Larson,** Wandering vectors for unitary systems and orthogonal wavelets, 1998

639 **Joan C. Artés, Robert E. Kooij, and Jaume Llibre,** Structurally stable quadratic vector fields, 1998

638 **Gunnar Fløystad,** Higher initial ideals of homogeneous ideals, 1998

637 **Thomáš Gedeon,** Cyclic feedback systems, 1998

636 **Ching-Chau Yu,** Nonlinear eigenvalues and analytic-hypoellipticity, 1998

635 **Magdy Assem,** On stability and endoscopic transfer of unipotent orbital integrals on p-adic symplectic groups, 1998

634 **Darrin D. Frey,** Conjugacy of Alt_5 and $\mathrm{SL}(2,5)$ subgroups of $\mathrm{E}_8(\mathbb{C})$, 1998

633 **Dikran Dikranjan and Dmitri Shakhmatov,** Algebraic structure of pseudocompact groups, 1998

632 **Shouchuan Hu and Nikolaos S. Papageorgiou,** Time-dependent subdifferential evolution inclusions and optimal control, 1998

631 **Ronnie Lee, Steven H. Weintraub, and J. William Hoffman,** The Siegel modular variety of degree two and level four/Cohomology of the Siegel modular group of degree two and level four, 1998

630 **Florin Rădulescu,** The Γ-equivariant form of the Berezin quantization of the upper half plane, 1998

629 **Richard B. Sowers,** Short-time geometry of random heat kernels, 1998

628 **Christopher K. McCord, Kenneth R. Meyer, and Quidong Wang,** The integral manifolds of the three body problem, 1998

627 **Roland Speicher,** Combinatorial theory of the free product with amalgamation and operator-valued free probability theory, 1998

626 **Mikhail Borovoi,** Abelian Galois cohomology of reductive groups, 1998

625 **George Xian-Zhi Yuan,** The study of minimax inequalities and applications to economies and variational inequalities, 1998

624 **P. Deift and K. T-R McLaughlin,** A continuum limit of the Toda lattice, 1998

623 **S. A. Adeleke and Peter M. Neumann,** Relations related to betweenness: Their structure and automorphisms, 1998

622 **Luigi Fontana, Steven G. Krantz, and Marco M. Peloso,** Hodge theory in the Sobolev topology for the de Rham complex, 1998

621 **Gregory L. Cherlin,** The classification of countable homogeneous directed graphs and countable homogeneous n-tournaments, 1998

620 **Victor Guba and Mark Sapir,** Diagram groups, 1997

619 **Kazuyoshi Kiyohara,** Two classes of Riemannian manifolds whose geodesic flows are integrable, 1997

618 **Karl H. Hofmann and Wolfgang A. F. Ruppert,** Lie groups and subsemigroups with surjective exponential function, 1997

(*Continued in the back of this publication*)

Controllability, Stabilization, and the Regulator Problem for Random Differential Systems

MEMOIRS
of the
American Mathematical Society

Number 646

Controllability, Stabilization, and the Regulator Problem for Random Differential Systems

Russell Johnson
Mahesh Nerurkar

November 1998 • Volume 136 • Number 646 (first of 6 numbers) • ISSN 0065-9266

American Mathematical Society
Providence, Rhode Island

1991 *Mathematics Subject Classification.*
Primary 34C35, 93D15.

Library of Congress Cataloging-in-Publication Data

Johnson, R. (Russell)
Controllability, stabilization, and the regulator problem for random differential systems
Russell Johnson, Mahesh Nerurkar.
p. cm. — (Memoirs of the American Mathematical Society, ISSN 0065-9266 ; no. 646)
"November 1998, volume 136, number 646 (first of 6 numbers)."
Includes bibliographical references.
ISBN 0-8218-0865-6 (alk. paper)
1. Differentiable dynamical systems. 2. Control theory. 3. Stability. I. Nerurkar, M. G. (Mahesh G.), 1952– . II. Title. III. Series.
QA3.A57 no. 646
[QA614.8]
510 s—dc21
[515′.35] 98-35279
CIP

Memoirs of the American Mathematical Society

This journal is devoted entirely to research in pure and applied mathematics.

Subscription information. The 1998 subscription begins with volume 131 and consists of six mailings, each containing one or more numbers. Subscription prices for 1998 are $435 list, $348 institutional member. A late charge of 10% of the subscription price will be imposed on orders received from nonmembers after January 1 of the subscription year. Subscribers outside the United States and India must pay a postage surcharge of $30; subscribers in India must pay a postage surcharge of $43. Expedited delivery to destinations in North America $35; elsewhere $110. Each number may be ordered separately; *please specify number* when ordering an individual number. For prices and titles of recently released numbers, see the New Publications sections of the *Notices of the American Mathematical Society.*

Back number information. For back issues see the *AMS Catalog of Publications.*

Subscriptions and orders should be addressed to the American Mathematical Society, P. O. Box 5904, Boston, MA 02206-5904. *All orders must be accompanied by payment.* Other correspondence should be addressed to Box 6248, Providence, RI 02940-6248.

Memoirs of the American Mathematical Society is published bimonthly (each volume consisting usually of more than one number) by the American Mathematical Society at 201 Charles Street, Providence, RI 02904-2294. Periodicals postage paid at Providence, RI. Postmaster: Send address changes to Memoirs, American Mathematical Society, P. O. Box 6248, Providence, RI 02940-6248.

This publication is indexed in *Science Citation Index*®, *SciSearch*®, *Research Alert*®, *CompuMath Citation Index*®, *Current Contents*®*/Physical, Chemical & Earth Sciences.*
Printed in the United States of America.

∞ The paper used in this book is acid-free and falls within the guidelines established to ensure permanence and durability.
Visit the AMS home page at URL: `http://www.ams.org/`

10 9 8 7 6 5 4 3 2 1 03 02 01 00 99 98

Contents

Abstract. We study in a dynamical systems context controllability and the random feedback stabilization problem for random control processes. We are led to study the random linear regulator problem, which we solve by considering the spectral theory of linear time-dependent Hamiltonian systems. This is done with the aid of the concepts of exponential dichotomy (hyperbolicity) and rotation number.

KEY WORDS: Dynamical Systems, Controllability, Stabilization.

A.M.S. Subject Classification: 34 C 35, 93 D 15

§0. Introduction.

Differential equations is a very classical subject and has been extensively studied over the past three hundred years. The qualitative theory of flows arising from autonomous (time-independent) systems is a vast and well-known area of mathematics. However comparatively little is known when the coefficients of the system depend on time. Moreover the intuition developed from the results in autonomous case is simply deceiving in the non-autonomous case. This is well-illustrated by the following example of Markus and Yamabe. Consider a 2×2 non-autonomous linear system;

$$x' = \begin{pmatrix} -2 + 2cos^2(t) & 1 - sin(2t) \\ -1 - sin(2t) & -2 + 2sin^2(t) \end{pmatrix} x.$$

Notice that for every value of $t \in \mathbf{R}$, the coefficient matrix $A(t)$ has eigenvalue $\lambda = -1$ with multiplicity two. However the origin is unstable because this equation admits an unbounded solution of the form $x(t) = \begin{pmatrix} -e^t cos(t) \\ e^t sin(t) \end{pmatrix}$. One natural question that arises about such systems concern their structural stability : if the coefficients are perturbed slightly, one wishes to know whether or not the structure of the solutions also changes slightly. Another related question concerns smoothness properties of the solutions as coefficient parameters are varied. Still another question concerns the solutions when "recurrence" or "randomness" is present in the coefficients. In this case, frequently one looks for solutions which exhibit the same degree of recurrence or randomness found in the coefficients.

It is therefore not surprising to find that non-autonomous control systems are not well understood, because after all they are frequently modelled by non-autonomous differential equations. A healthy body of results exists for linear and non-linear autonomous control processes. However there is perhaps a lack of awareness in the control community (and elsewhere) that many standard results for autonomous systems are not only false in the non-autonomous setting, but do not provide an appropriate intuition for approaching such systems. The main purpose of the present article is to demonstrate this assertion by developing a systematic study of time-dependent control processes. The subject of control theory and optimization is of course extremely vast. We shall limit ourselves to the basic problem of null controllability of linear systems, and to the widely studied problem of feedback stabilization of linear systems. This in turn involves the study of the well known optimization problem of the linear regulator with quadratic cost functionals. The notion of uniform controllability plays an important role in the solutions of these problems. Finally we shall extend our discussion and results to non-linear situations via the technique of linearization in the non-autonomous and random set-up.

We will study control processes whose coefficients are **random**. These make up a very large subset of non-autonomous systems. By random we will mean systems with the whole range of behaviour from periodic coefficients to coefficients which are (bounded) independent, identically distributed stochastic processes. Stochastic methods, powerful as they are, can be only applied when the coefficients exhibit very strong independence properties. Our methods apply when such independence is not present (and, often, when it is present).

Russell Johnson' s research was supported by the M. U. R. S. T. (Italy)
Received by the editor February 12, 1996.

We will study random processes using methods of dynamical systems. The main point to underline is that, if the system coefficients are non- periodic, then the study of solution behaviour is rendered very difficult by **recurrence** properties of the coefficients. This complicated qualitative behaviour is already illustrated by the almost periodic systems. This complication arises because of the absence of a Floquet matrix which in turn is due to the **non-periodic recurrence** in such systems. The natural method for studying recurrent systems is the use of ergodic theory and dynamical systems-the method that we will use. In fact over the past decade certain developments in mathematical physics (particularly in the theory of random Schrödinger operators) have well demonstrated the sensitive dependence of the spectral properties of operators on the non-periodic recurrence properties of their potential functions. Our methods are in fact motivated by their success in the study of the random Schrödinger operator.

In this article, we will show how difficulties due to non-periodic recurrence can be overcome and in fact can be used to obtain results which are counter intuitive from an autonomous point of view. For example, we shall see that in the non-autonomous linear case, the system can have positive Lyapunov exponents and can still be globally null controllable. Also we shall see examples of control processes with quasi-periodic coefficients where the structure of the reachable set is extremely sensitive to the continued fraction properties of its frequencies. In addition recurrence results from ergodic theory will allow us to design controls which steer sub- exponentially drifting trajectories to zero in finite time. Finally, methods of dynamical systems will allow us to solve the random linear feedback control problem. We will obtain a solution of the Riccati equation with extremely good robustness properties, which also preserves all smoothness and recurrence properties of the coefficients.

We will see that the non-autonomous nature of the coefficients of the system gives rise to a natural 'skew-product' structure on the space of its solutions ([MS]; see Section 3, below). Intuitively one can think of this skew-product structure in the following way. On the one hand, the coefficients exhibit dynamics which is independent of those state variables $x_1, \cdots, x_n$. On the other hand, a given state variable x_i evolves in a way which depends on its interaction with the other state variables and on the dynamics of the coefficients.

§1. Basic Dynamical Notions.

We shall be discussing differential equations and control processes whose coefficient functions are time dependent or "random". These functions may exhibit the entire range of behaviour from time periodic or almost periodic to highly stochastic behaviour with positive entropy and also the intermediate case of zero entropy with or without mixing conditions. We use the word "random" to refer to this very wide class of functions. Such a wide class of functions can be best modelled by thinking of them as functions evaluated along a trajectory of a dynamical system where the dynamical system is suitably chosen to reflect the random behaviour of the coefficient functions. Thus in practice our dynamical system will be the "**hull**" of the coefficient functions. We begin now with the formal definitions.

1.1 Definition. *A* **dynamical system** *(or a* **flow***) is a pair* $(\Omega, \{T_t\}_{t \in \mathbf{R}})$ *where* Ω *is a compact metric space and* $T_t : \Omega \to \Omega$, $(t \in \mathbf{R})$ *is a one parameter group of homeomorphisms of* Ω *in the sense that*

(1) $T_0 = I$*-the identity map,*
(2) $T_{t+s} = T_t \circ T_s$, $\quad (t, s \in \mathbf{R})$ *and*
(3) *the map* $(\omega, t) \to T_t(\omega)$ *is jointly continuous.*

1.2 Definition. *(a) Given* $\omega \in \Omega$, *the* **orbit of** ω *is the set*

$$Orb(\omega) = \{T_t(\omega) \mid t \in \mathbf{R}\}.$$

A subset $M \subseteq \Omega$ *is* **invariant** *if* $Orb(m) \subseteq M$ *for all* $m \in M$. *A system* $(\Omega, \{T_t\}_{t \in \mathbf{R}})$ *is* **minimal** *if and only if there are no proper closed invariant subsets or equivalently each orbit is dense.*
(b) The set M *is* **chain recurrent** *if given numbers* $\varepsilon > 0, T > 0$ *and points* $\omega, \overline{\omega} \in M$, *there exists points* $\omega = \omega_0, \omega_1, \cdots, \omega_n = \overline{\omega}$ *and numbers* $t_0 > T, \cdots, t_{n-1} > T$ *such that*

$$d(T_{t_i}(\omega_i), \omega_{i+1}) < \varepsilon, \qquad (0 \leq i \leq n-1).$$

Here d *is some metric on* Ω *generating the topology on* Ω.

It is easy to see that Ω always contains a minimal set (see [E]). Also minimality implies chain recurrence.

1.3 Definition. *Let* μ *be a Radon probability measure on* Ω.
(1) Then μ *is* **invariant** *if* $\mu(T_t(B)) = \mu(B)$ *for each Borel subset* $B \subseteq \Omega$ *and for each* $t \in \mathbf{R}$.
(2) An invariant measure μ *is said to be* **ergodic** *if for each Borel subset* $B \subseteq \Omega$ *the condition* $\mu(B \Delta T_t(B)) = 0$ *for all* $t \in \mathbf{R}$ *implies either* $\mu(B) = 0$ *or* $\mu(B) = 1$, *(*Δ *denotes the symmetric difference of sets).*
(3) Finally, the **topological support** *of* μ *is defined by the complement of the largest open subset* $V \subseteq \Omega$ *for which* $\mu(V) = 0$.

1.4 Lemma. *Suppose that* Ω *is the topological support of an ergodic measure* μ. *Then* Ω *is chain recurrent.*

Proof. It suffices to prove that Ω contains a point ω_0 whose positive semi-orbit, i.e the set $\{T_t(\omega_0) \mid t > 0\}$ is dense in Ω. Let $\{V_i \mid i \in \mathbf{N}\}$ be a basis of open subsets of Ω. Then $\mu(V_i) > 0$ for each $i \in \mathbf{N}$. By the Birkhoff ergodic theorem, the set $Y_i = \{\omega \in \Omega \mid T_t(\omega) \in V_i \text{ for some } t > 0\}$ has μ measure 1. Hence the set $\cap\{Y_i \mid i \in \mathbf{N}\}$ has also μ measure 1. Now note that each point in this intersection has dense positive semi- orbit. □

§2. Random Linear Control Processes.

First we consider linear control systems modelled over a given flow $(\Omega, \{T_t\}_{t \in \mathbf{R}})$. Let $M(n,m)$ be the set of $n \times m$ real matrices with the Euclidean norm $|| \ ||$. Let $A : \Omega \to M(n,m)$ be a measurable map satisfying the following two conditions.
(a) The maps $t \to ||A(T_t(\omega))||^p$ are uniformly integrable, that is

$$\sup_{\omega} \sup_{t} \int_t^{t+1} ||A(T_t(\omega))||^p ds < \infty.$$

If $p = 1$, suppose in addition that

$$\lim_{\varepsilon \to 0} \int_t^{t+\varepsilon} ||A(T_s(\omega))|| ds = 0$$

uniformly in $t \in \mathbf{R}$ and $\omega \in \Omega$.
(b) For any smooth function $\varphi : \mathbf{R} \to \mathbf{R}^n$ of compact support, the following map

$$\omega \to \int_{\mathbf{R}} A(T_t(\omega))\varphi(t) dt$$

is continuous.

Now let $B : \Omega \to M(n,m)$ be another map that is continuous. Consider the following family of linear control processes parametrized by a point $\omega \in \Omega$;

$$(2.1.\omega) \qquad x' = A(T_t(\omega))x + B(T_t(\omega))u, \qquad (\omega \in \Omega),$$

where $u = u(t)$ is an appropriate control function. Our control functions will be locally integrable and we shall impose a magnitude constraint on u, that is $u(t)$ is required to lie in a fixed compact convex subset $\Lambda \subseteq \mathbf{R}^m$ containing the origin. We shall refer to such controls as **admissible controls**.

We demonstrate how the above set up includes non-autonomous systems with a very wide class of coefficient functions. For example, consider the linear non-autonomous control process given by;

$$(2.1) \qquad x'(t) = a(t)x(t) + b(t)u(t) \qquad (x \in \mathbf{R}^n, u \in \mathbf{R}^m)$$

where a and b respectively belong to class $\mathcal{C}$ and $\mathcal{D}$ of functions defined as follows. Fix a number $1 \leq p \leq \infty$, and let

$$\mathcal{C} = \{a : \mathbf{R} \to M(n,n) \mid \sup_{t \in \mathbf{R}} \int_t^{t+1} ||a(s)||^p ds < \infty\} \text{ if } p < \infty,$$
$$\mathcal{C} = L^\infty(\mathbf{R}, M(n,n)) \text{ if } p = \infty.$$

If $p = 1$, then in addition suppose that

$$\lim_{\varepsilon \to 0} \int_t^{t+\varepsilon} ||a(s)|| ds = 0$$

uniformly in $t \in \mathbf{R}$. Give $\mathcal{C}$ the distribution topology; that is $a_n \to a$ in $\mathcal{C}$ if and only if

$$\int_{\mathbf{R}} a_n(t)\varphi(t)dt \to \int_{\mathbf{R}} a(t)\varphi(t)dt$$

for each smooth function $\varphi : \mathbf{R} \to \mathbf{R}^n$ of compact support. Next define

$$\mathcal{D} = \{b : \mathbf{R} \to M(n,m) \mid b \text{ is uniformly bounded and uniformly continuous}\}.$$

Give $\mathcal{D}$ the topology of uniform convergence on compact sets. Both $\mathcal{C}$ and $\mathcal{D}$ support the **translation flow** $\{T_t\}_{t\in\mathbf{R}}$, where T_t is the translation:

$$(T_t a)(s) = a(t+s), \quad (T_t b)(s) = b(t+s) \qquad (t, s \in \mathbf{R}).$$

We will choose the coefficient functions $a(t), b(t)$ in (2.1) from sets $\mathcal{C}$ and $\mathcal{D}$ respectively. Let Ω be a compact, translation invariant subset of $\mathcal{C} \times \mathcal{D}$. Thus in particular Ω is metrizable.

If $\omega = (a,b) \in \Omega$, define $B(\omega) = b(0)$; then it is easily seen that $B : \Omega \to M(n,m)$ is continuous and $B(T_t(\omega)) = b(t)$, $(t \in \mathbf{R})$. On the other hand, while there does exist a Borel function $A : \Omega \to M(n,n)$ such that, for each $\omega = (a,b) \in \Omega$, the equation $A(T_t(\omega)) = a(t)$ holds for almost all $t \in \mathbf{R}$, the function A may not be continuous. For our purposes we can agree to regard the expression "$t \to A(T_t(\omega))$" as a notational device for expressing the function $a(t)$. That is $t \to A(T_t(\omega))$ represents the projection of ω onto its first co-ordinate.

Thus our point of view is quite general in that we do not require the existence of a point $\omega_0 \in \Omega$ whose orbit is dense in Ω. This framework includes all time-varying systems with bounded coefficients, from periodic and almost periodic to highly stochastic systems with positive entropy, also the intermediate case of zero entropy with or without mixing conditions. We use the word "random" to refer to this very wide class of coefficients.

We turn now to control theory, and recall the concept of local and global null controllability. We define these concepts for a single control process of the form (2.1). In the case of a random system (2.1.ω), these notions will apply to every single process corresponding to a specific choice of $\omega \in \Omega$.

2.1 Definition. *Let $(x_0, t_0) \in \mathbf{R}^n \times \mathbf{R}$. We say that (x_0, t_0) can be ω-***steered to** *y in time $T > 0$ if there exists an admissible control $u_0 : [t_0, t_0+T] \to \Lambda$ such that the solution of the initial value problem (2.1.ω) satisfying $x(t_0) = x_0$ also satisfies $x(t_0+T) = y$. Most of the time y will be the origin in $\mathbf{R}^n$ and t_0 will be equal to zero; in this case we say that x_0 can be ω-steered to $y = 0$ in time T. Set*

$$D(\omega, T) = \{x \in \mathbf{R}^n \mid x \textit{ can be } \omega \textit{ steered to } y = 0 \textit{ in time } T\} \quad \textit{and}$$

$$D(\omega) = \cup\{D(\omega, T) \mid T > 0\} \qquad \textit{for each } \omega \in \Omega. \tag{2.2}$$

With this notation,

(1) *the system* (2.1.ω) *is* **locally null controllable** *at* ω *if there exists* $T > 0$ *and a neighbourhood* V *of* $0 \in \mathbf{R}^n$ *such that* $V \subseteq D(\omega, T)$.

Note that one can define a weaker notion of local null controllability by letting T *depend on* $x \in V$. *However, these notions turn out to be the same under mild assumptions on* Λ *(see Corollary 4.3)).*

(2) *The control process* (2.1.ω) *is called* **globally null controllable** *at* ω *if* $D(\omega) = \mathbf{R}^n$.

(3) *Finally, the system of control processes* (2.1.ω) *is said to be* **uniformly locally null controllable** *if there exists a neighbourhood* V *of the origin in* $\mathbf{R}^n$ *and a number* $T > 0$ *such that* $V \subseteq D(\omega, T)$ *for all* $\omega \in \Omega$.

Now we briefly discuss those aspects of controllability which we shall study in some detail. We shall consider sufficiency conditions for uniform null controllability. We shall prove that if a system is controllable for some ω in each minimal subset of Ω, then it is in fact uniformly null controllable. This is a far reaching generalization of a result of Z. Artstein [As]. In the random case the nature of the reachable set is very sensitive to the nature of the flow on Ω. We illustrate this with a constructive example employing the ideas of dynamical systems. Finally we address the question: when does local null controllability implies global null controllability? Here we shall prove some counter intuitive results which are in fact false in the constant coefficient case. These surprising results stem from the complicated recurrence behaviour of the coefficients. We shall use recurrence results from ergodic theory and the spectral theory developed by Sacker and Sell [SS] to analyze the complications caused by the non-periodic recurrence.

Our methods in the subsequent chapters will rely on the qualitative theory of random linear differential systems. Hence we pause here to first describe that theory and the relevant results.

§3. Some Facts About Random Linear Systems.

Consider the linear differential system associated to the control process (2.1.ω)

$$x' = A(T_t(\omega))x, \qquad (x \in \mathbf{R}^n, \omega \in \Omega). \tag{3.1.ω}$$

Let $X_A(\omega, t)$ be the fundamental matrix solution of the above equation satisfying the initial condition $X_A(\omega, 0) = I$, where I is the identity matrix. Then the map $X_A : \Omega \times \mathbf{R} \to GL(n, \mathbf{R})$ is jointly continuous and satisfies the following **cocycle identity**:

$$X_A(\omega, t+s) = X_A(T_t(\omega), s) X_A(\omega, t) \qquad (\omega \in \Omega, t, s \in \mathbf{R}). \tag{3.2}$$

We will also need to consider the **adjoint system**

$$x' = -A^*(T_t(\omega))x \qquad (\omega \in \Omega). \tag{3.3.ω}$$

(here $*$ denotes the transpose). The fundamental matrix solution of this system is given by $Z_A(\omega, t) = X_A^*(\omega, t)^{-1}$.

The qualitative behaviour of solutions of the linear system is studied through its "polar co-ordinates". This consists of the study of the "angular behaviour" and the study of the "growth rate" of its norm. The angular behaviour is best captured by the dynamics of the **skew product flow** defined on the projective bundle $\mathbf{P} \times \Omega$, $\mathbf{P}$ being the $n-1$ dimensional projective space. This flow is defined as follows:

$$T_t^A([v], \omega) = ([X_A(\omega, t)v], T_t(\omega)), \qquad (([v], \omega) \in \mathbf{P} \times \Omega), \tag{3.4}$$

where $[v]$ denotes the ray through the vector $v \in \mathbf{R}^n$. Note that the same formula also defines a flow on the vector bundle $\mathbf{R}^n \times \Omega$.

The radial component of the polar co-ordinate is studied via what is commonly known as the Lyapunov exponents.

3.1 Definition. *Let* $([v], \omega) \in \mathbf{R}^n \times \Omega$. *The numbers*

$$\beta(v, \omega) = \limsup_{t \to \infty} \frac{1}{t} \, ln \, ||X_A(\omega, t)v||,$$

are called the **Lyapunov exponents** *of the cocycle. It is easily seen that for each fixed* ω *there are only finitely many distinct Lyapunov exponents.*

The Lyapunov exponents can be expressed as ergodic averages of a suitable function with respect to the skew-product flow on $\mathbf{P} \times \Omega$. This function is given by

$$H([v], \omega) = \frac{\langle A(\omega)v, v \rangle}{||v||^2}.$$

Then it is easy to observe that

$$\beta(v, \omega) = \limsup_{t \to \infty} \frac{1}{t} \int_0^t H(T_s^A([v], \omega)) ds.$$

If μ is an invariant ergodic measure for the flow $(\Omega, \{T_t\}_{t \in \mathbf{R}})$, then for μ a.e. $\omega \in \Omega$ the set of Lyapunov exponents are independent of ω [JPS]. We write $\{\beta_i(\mu) \mid 1 \leq i \leq s\}$ for the set of distinct Lyapunov exponents which are independent of ω for μ a.e. $\omega \in \Omega$.

The stability theory of linear (time dependent) vector fields will play a central role in our results. The key idea in the study of stability is hyperbolicity. We begin with the definition of exponential dichotomy which is exactly the notion of hyperbolicity in the time dependent case.

3.2 Definition. *The cocycle* $X_A(\omega, t)$ *is said to be* **hyperbolic** *or equivalently is said to admit an* **exponential dichotomy** *(ED for short) over* Ω *if there exists continuous vector subbundles* V^s, V^u *of the vector bundle* $\mathbf{R}^n \times \Omega$ *satisfying the following properties:*

1. $V^s \oplus V^u = \mathbf{R}^n \times \Omega$ *(the Whitney sum);*
2. V^s *and* V^u *are invariant under the skew product flow* $\{T_t^A\}_{t \in \mathbf{R}}$ *on* $\mathbf{R}^n \times \Omega$;
3. *there are constants* $K > 0$ *and* $\alpha > 0$ *such that,*

$$\text{if } (v, \omega) \in V^s \text{ then } ||X_A(\omega, t)v|| \leq K e^{-\alpha t} ||v|| \qquad (t \geq 0) \text{ and}$$
$$\text{if } (v, \omega) \in V^u \text{ then } ||X_A(\omega, t)v|| \leq K e^{\alpha t} ||v|| \qquad (t \leq 0).$$

For later use, we introduce a continuous family of projection operators $\omega \to P(\omega)$, *where* $P(\omega)$, $(\omega \in \Omega)$ *is the projection onto the stable subspace* $V^s(\omega)$ *in the fiber over* ω.

Several different criteria for checking hyperbolicity are known (see [Cp],[MSc]). In the autonomous case ED is equivalent to the requirement that the real part of the eigenvalues of the (constant) matrix A are non-zero or equivalently the set of eigenvalues of e^A is disjoint from the unit circle. For the tangent cocycle of an Anosov diffeomorphism having a dense set of periodic points, an analogue of above characterization was proved by J. Mather as well as by R. Mane. This was then generalized to hyperbolic cocycles independently by C. Chicone and R. Swanson [CS] and R. Johnson [J3]. These results play a part in non-autonomous (random) linearization of nonlinear systems.

There is another important criterion for checking hyperbolicity. The following theorem due to Selgrade [Sg] describes it.

3.3 Theorem. *Suppose the flow* $(\Omega, \{T_t\}_{t\in\mathbf{R}})$ *is chain recurrent. Then the cocycle* $X_A(\omega, t)$ *admits an ED over* Ω *if and only the equations* $(3.1.\omega)$ *does not admit any non-zero bounded solution for all* $\omega \in \Omega$.

Next, we define the notion of dichotomy spectrum of a cocycle.

3.4 Definition. *The* **dichotomy spectrum** *of the cocycle* $X_A(\omega, t)$ *is the set*

$$\Sigma \equiv \Sigma(X_A) = \{\lambda \in \mathbf{R} \mid e^{-\lambda t}X(\omega, t) \text{ is not hyperbolic}\}.$$

When $A(\omega) \equiv A$ is a constant function, Σ is just the set of real parts of eigenvalues of A. The following spectral theorem of Sacker and Sell [SS] shows that the dichotomy spectrum is a finite union of intervals.

3.5 Theorem. *Let* X_A *be a cocycle in to* $GL(n, \mathbf{R})$. *Then its dichotomy spectrum* Σ *is a disjoint union of* k $(k \le n)$ *compact intervals* $[a_i, b_i] \subseteq \mathbf{R}$, $1 \le i \le k$. *Furthermore, there exist* k *continuous subbundles* $W_1, W_2, \cdots, W_k$ *of the trivial bundle* $\mathbf{R}^n \times \Omega$ *such that:*

(1) $\mathbf{R}^n \times \Omega = W_1 \oplus W_2 \oplus \cdots \oplus W_k$ *(Whitney sum);*
(2) *each* W_i $(1 \le i \le k)$ *is invariant under the skew product flow* $(\mathbf{R}^n \times \Omega, \{T_t^A\}_{t\in\mathbf{R}})$;
(3) $\limsup_{t\to \pm\infty} \frac{1}{t}\, ln||X_A(\omega, t)v|| \in [a_i, b_i]$ *and* $\liminf_{t\to \pm\infty} \frac{1}{t}\, ln||X_A(\omega, t)v|| \in [a_i, b_i]$;
(4) *the spectrum of the cocycle restricted to the invariant subbundle* W_i *is* $[a_i, b_i]$, *for each* $1 \le i \le k$.

Analogously one defines the spectrum Σ^* of the adjoint system $(3.3.\omega)$. It is easily seen that $\Sigma = -\Sigma^*$; that is $\lambda \in \Sigma$ if and only if $-\lambda \in \Sigma^*$.

The complexity of the qualitative behaviour is reflected in the non- degeneracy of the spectral intervals, (i.e. when the spectral intervals are not singleton sets). Of course in the autonomous case (and by the Floquet theorem in the periodic case as well) the spectral intervals are always degenerate. We shall see that when such non-degenerate spectral intervals exist, one gets surprising results for the corresponding control system (e.g. see Theorem 6.3).

The following proposition shows that every Lyapunov exponent is in a spectral interval and the end points of each spectral interval are Lyapunov exponents, [JPS].

3.6 Proposition. *(a) Let μ be an invariant ergodic measure on Ω. Then every Lyapunov exponent $\beta_i(\mu)$ of the cocyle generated by (3.1.ω) belongs to the dichotomy spectrum.*
(b) If $\beta = c_j$ or $\beta = d_j$ is an end point of a spectral interval $[c_j, d_j] \subseteq \Sigma$, then there exists an ergodic measure μ on Ω and a corresponding Lyapunov exponent $\beta_i(\mu)$ such that $\beta = \beta_i(\mu)$.

We need another technical ingredient from the spectral theory of random linear systems, namely the technique of **recurrent triangularization** of a linear cocycle. First we review the basic construction (see [JPS]). Let $Y = O(n) \times \Omega$ where $O(n)$ is the orthogonal group of $\mathbf{R}^n$. We define a flow on Y by using the Gram-Schmidt orthogonalization procedure as follows.

As before, let $X_A(\omega, t)$ be the cocycle generated by (3.1.ω). Let $\omega \in \Omega$, $g_0 \in O(n)$, and write $y = (g_0, \omega) \in Y$. By the Gram-Schmidt process, we obtain

$$X_A(\omega, t)g_0 = \Gamma(y, t)R(y, t) = \Gamma(g_0, \omega, t)R(g_0, \omega, t),$$

where $\Gamma(y, t) \in O(n)$ and $R(y, t)$ is a triangular matrix with zeroes above the main diagonal and positive diagonal entries. Now using the cocycle identity (3.2) and the uniqueness in the Gram-Schmidt decomposition, one can show that (see [JPS])

$$T_t^A(g_0, \omega) = (\Gamma(y, t), T_t(\omega))$$

defines a flow on Y. Furthermore, with respect to this flow, the map $R(y, t)$ satisfies the cocycle identity;

$$R(y, t + s) = R(T_t^A(y), s)R(y, t) \qquad (t, s \in \mathbf{R}).$$

Let M be a minimal set of Y. Consider the restriction of the cocycle R to $M \times \mathbf{R}$. We write

$$r(m) = \frac{d}{dt} R(T_t^A(m))\big|_{t=0},$$

so that $R(m, t)$ is the fundamental matrix solution to the differential equation

$$w' = r(T_t^A(m))w \tag{3.5.m}$$

which satisfies $R(m, 0) = I$. We recall that $r(T_t^A(m)) = [r_{i,j}(T_t^A(m))]$ is a matrix whose entries above the main diagonal are equal to zero.

There is a close relationship between equations (3.1.ω) and equations (3.5.m). Namely, we can lift equations (3.1.ω) to M by introducing the natural projection $\pi : M \to \Omega,\ \pi(\omega, g) = \omega$ and defining

$$\hat{A}(m) = A \circ \pi(m) \quad (m \in M).$$

Consider the lift of cocycle $X_A(\omega, t)$ to M, namely set, $X_{\hat{A}}(m,t) = X_A(\pi(m), t) = X_A(\omega, t)$. Then the orthogonal change of variables

$$x = F(T_t^A(m))w, \tag{3.6}$$

where $F : Y \to O(n),\ F(y) = g$ for $y = (g, \omega) \in Y$, transforms equations (3.1.ω) (lifted to M via π) in to equations (3.5.m). In fact we have the following "cohomology" relation

$$R(m,t) = F(T_t^A(m))X_A(\omega,t)F(m)^{-1} \qquad (\pi(m) = \omega,\ m \in M).$$

Thus, up to a lift to M, the cocycle X_A is "cohomologous" to a cocycle in the group of triangular matrices. This is called the "recurrent triangularization" of the cocyle X_A.

Now we introduce the notion of irreducibility.

3.7 Definition. *First we introduce the numbers,*

$$\rho^-(i) = \inf_{m \in M} \liminf_{t \to \infty} \frac{1}{t} \int_0^t r_{i,i}(T_t^A(m))ds,$$

$$\rho^+(i) = \sup_{m \in M} \limsup_{t \to \infty} \frac{1}{t} \int_0^t r_{i,i}(T_t^A(m))ds,$$

where $1 \le i \le n$. *Let* J_i *be the interval* $[\rho^-(i), \rho^+(i)] \subset \mathbf{R}$, $(1 \le i \le n)$. *Equations* (3.1.ω) *are called* **irreducible** *if there exists a minimal subset* $M \subset Y = O(n) \times \Omega$ *such that the intervals* J_i *have a point in common; i.e. the set* $\cap\{J_i \mid 1 \le i \le n\}$ *is non-empty.*

To clarify the significance of irreducibility, let μ be an ergodic measure on Ω, and let $\hat{\mu}$ be a Borel probability measure on M that is ergodic with respect to the flow $(M, \{T_t^{\hat{A}}\}_{t \in \mathbf{R}})$ and is also a lift of μ; i.e. $\hat{\mu}(\pi^{-1}(B)) = \mu(B)$ for all Borel sets $B \subset \Omega$. Such a measure can always be found [JPS]. Then for $\hat{\mu}$ a.a. $m \in M$, the set of Lyapunov exponents of equations (3.5.m) equals the set of mean values

$$\int_M r_{i,i}(m)d\hat{\mu}(m) \quad (1 \le i \le n). \tag{3.7}$$

Hence for μ a.a. $\omega \in \Omega$, the set of Lyapunov exponents of (3.1.ω) equals the set of mean values given by the expression (3.7).

Now by the Birkhoff ergodic theorem:

$$\lim_{t \to \infty} \frac{1}{t} \int_0^t r_{i,i}(T_t^{\hat{A}}(m))ds = \int_M r_{i,i}(m)d\hat{\mu}(m) \quad \text{for } \hat{\mu} \text{ a.a. } m.$$

Thus, roughly speaking, each interval J_i is bounded by "lower" and "upper" exponents. More precisely, the following result can be proved [JPS].

3.8 Proposition. *For each $1 \leq i \leq n$, there exist ergodic measures $\hat{\mu}$, $\hat{\nu}$ on M such that $\rho^-(i) = \int_M r_{i,i}(m)d\hat{\mu}_i(m)$ and $\rho^+(i) = \int_M r_{i,i}(m)d\hat{\nu}_i(m)$.*

Letting μ_i, ν_ibe the projections of $\hat{\mu}_i, \hat{\nu}_i$ on Ω, we see that $\rho^{\pm}(i)$ are true almost everywhere exponents of equations (3.1.ω), though in general with respect to distinct ergodic measures on Ω. If Ω admits a **unique** ergodic measure μ (this is the case if, for example, $a(t)$ and $b(t)$ are Bohr almost periodic), then each value $\rho^{\pm}(i)$ is one of the exponents $\{\beta_i(\mu) \mid 1 \leq i \leq s\}$ corresponding to μ. It should be emphasized that M need not be uniquely ergodic even if Ω is uniquely ergodic.

We give an example to show that a family of equations (3.1.ω), irreducible in the sense of Definition 3.7 may actually be reducible in a natural sense. For this, let $A_1(t)$ be the 2×2 almost periodic function constructed by V. Millionschikov [M1]. The dichotomy spectrum of the cocycle corresponding to this matrix function is a non- degenerate interval $[-\beta, \beta]$, see [J2],[M1]. Define

$$A(t) = \begin{pmatrix} A_1(t) & 0 \\ 0 & a \end{pmatrix}$$

where a is a real number in $[-\beta, \beta]$. It can be shown that, after triangularization, two of the three intervals J_i are equal to $[-\beta, \beta]$ and the third reduces to a. On the other hand, the block form of $A(t)$ clearly induces a reduction of the space of solutions of $x' = A(t)x$ to a sum of a two-dimensional bundle and a one-dimensional bundle.

It is natural to ask about the relation between irreducibility and the properties of the dichotomy spectrum Σ. Using the technique of recurrent triangularization and Theorem 3.3 one obtains the following, see [JN4].

3.9 Proposition. *The dichotomy spectrum Σ equals $\cup\{J_i \mid 1 \leq i \leq n\}$.*

As an immediate corollary one obtains the following.

3.10 Corollary. *If equations (3.1.ω) are irreducible, then the dichotomy spectrum Σ is a single interval.*

We finish this section by considering systems of dimension $n = 2$ or 3.

3.11 Proposition. *Let the flow $(\Omega, \{T_t\}_{t\in\mathbf{R}})$ admit a unique ergodic measure μ. For example, this condition is satisfied by the translation flow on the hull of Bohr almost periodic functions.*
(a) If $n = 2$ and if the equations (3.1.ω) do not admit an exponential dichotomy, then J_1 and J_2 are both equal to the dichotomy spectrum Σ.
(b) If $n = 3$, then the equations (3.1.ω) are irreducible if and only if Σ is a single interval.

Proof. (a) This follows from the structure theory for two-dimensional systems presented in [J2]. The results given there are stated for almost periodic system, but the arguments needed to prove (a) carry over to the case when the flow $(\Omega, \{T_t\}_{t\in\mathbf{R}})$ is uniquely ergodic.
(b) We need prove only the 'if' statement. Let $C = \{\beta_1(\mu), \beta_2(\mu), \beta_3(\mu)\}$ be the set of almost everywhere Lyapunov exponents of equations (3.1.ω) with respect to μ. By

Proposition 3.8 and the discussion following it, we see that $\rho^{\pm}(i) \in C$ for each choice of $\pm$ and for $i = 1, 2, 3$. We can assume that $\beta_1(\mu) \le \beta_2(\mu) \le \beta_3(\mu)$. Now, we know that the union of the intervals J_i equals $\Sigma = [\beta_1(\mu), \beta_3(\mu)]$.

If C consists of three distint elements, then an argument using the construction of and the uniqueness of the "Oseledec bundles" [JPS] shows that there are two possibilities. One is that each of the three intervals J_i does not reduce to a point, in which case $\beta_2(\mu)$ is common to all of them. The other is that exactly one interval J_i reduces to a point, and this point is $\beta_2(\mu)$. Thus one always has $\beta_2(\mu) \in \cap\{J_i \mid 1 \le i \le 3\}$ if C contains three distinct points.

If C contains exactly two points $\beta_1(\mu) < \beta_2(\mu)$, then similar arguments show that at most one interval J_i can reduce to a point; that point must be either $\beta_1(\mu)$ or $\beta_2(\mu)$. Thus either $\beta_1(\mu)$ or $\beta_2(\mu)$ lies in $\cap\{J_i \mid 1 \le i \le 3\}$.

Finally, if C is a singleton $\{\beta\}$, then clearly $\beta \in \cap\{J_i \mid 1 \le i \le 3\}$. So we conclude that equations (3.1.ω) are irreducible. ☐

§4. Sufficiency Conditions for Uniform Controllability.

Consider the linear control process (2.1.ω)

$$x' = A(T_t(\omega))x + B(T_t(\omega))u, \qquad \omega \in \Omega,$$

with the constraint $u(t) \in \Lambda$, where $\Lambda \subseteq \mathbf{R}^m$ is a given constraint set. Here we shall consider the following question: What are the conditions under which local null controllability of (2.1.ω) for some $\omega \in \Omega$ implies (uniform) local null controllability for all $\omega \in \Omega$. To incorporate the constraints in to our analysis, following [BS1] we introduce the following notion of a support function.

4.1 Definition. *Let Λ be a subset of $\mathbf{R}^m$. Define the map $H_\Lambda : \mathbf{R}^m \to \mathbf{R}$ by setting*

$$H_\Lambda(\alpha) = Sup\ \{\langle \alpha, v\rangle \mid v \in \Lambda\}$$

where $\langle\ ,\ \rangle$ is the usual inner product on $\mathbf{R}^m$. The map H_Λ is called the **support function** *of set Λ.*

Notice that, if $0 \in \Lambda$ then H_Λ is non-negative and if Λ is compact, then H_Λ is continuous. First, we recall a necessary and sufficient condition for local null controllability due to B. Barmish and W. Schmitendorf [BS1,2]. This lemma is valid for the linear control process (2.1) solely under the condition that $a(t)$ and $b(t)$ are locally integrable.

4.2 Lemma. *Consider the control process* (2.1)

$$x'(t) = a(t)x(t) + b(t)u(t), \qquad x \in \mathbf{R}^n,\ u \in \mathbf{R}^m,$$

with the constraint $u(t) \in \Lambda$, where Λ is compact and contains the origin. Let $x_0 \in \mathbf{R}^n$ and $t_0 \in \mathbf{R}$.

(a) The point (x_0, t_0) *can be steered to zero in time* T *if and only if*

$$\langle x_0, z^*(t_0)\rangle + \int_{t_0}^{t_0+T} H_\Lambda(B^*(s)z^*(s))ds \geq 0$$

for all solutions $z^*(t)$ *of the associated adjoint linear system.*
(b) The process is locally null controllable if and only if there exists $\epsilon > 0$ *such that*

$$\int_0^\infty H_\Lambda(B^*(s)z^*(s))ds \geq \epsilon$$

for each solution $z^*(t)$ *of the associated adjoint linear system that satisfies the initial condition* $||z^*(0)|| = 1$, *(here* $||\cdot||$ *is the Euclidean norm on* $\mathbf{R}^n$*).*

4.3 Corollary. *Consider the control process* (2.1) *with the constraints set* Λ*. Suppose there is a neighbourhood* V *of the origin in* $\mathbf{R}^n$ *such that for each* $x \in V$*, there exists* $T \equiv T(x) > 0$ *with the property that* x *can be steered to the origin in time* T*. Then the control process* (2.1) *is locally null controllable if either (a)* Λ *is compact, convex and contains the origin or (b)* Λ *is compact and* $0 \in \Lambda^0$*-the interior of* Λ*.*

Proof. (a) Choose $x_1, x_2, \cdots, x_r \in V$ such that the convex hull of the x_i's contains a ball B centered at the origin. Let u_i, $1 \leq i \leq r$ be an admissible controls that steers x_i to the origin in time T_i. Set $T = \text{Max}\ \{T_i \mid 1 \leq i \leq r\}$. If $x \in B$ then $x = \sum_{i=1}^r \epsilon_i x_i$ where $0 \leq \epsilon_i \leq 1$. Define $u = \sum_{i=1}^r \epsilon_i \tilde{u_i}$ where $\tilde{u_i} = u_i$ on $[0, T_i]$ and zero otherwise. Then u steers x to the origin in time T. Hence the same T can be used for all points in B and part (a) is proved.
(b) Let $\tilde{\Lambda}$ be the convex hull of Λ. Let δ and R be positive numbers chosen so that (i) the ball $B(\delta)$ centered at the origin with radius δ is contained in Λ and (ii) the ball $B(R)$ centered at the origin with radius R contains $\tilde{\Lambda}$. Now applying part (a) of Lemma 4.2 to the constrain set $B(R)$ we obtain an $\epsilon > 0$ such that, every solution $z^*(t)$ of the adjoint linear system satisfying $||z^*(0)|| = 1$ also satisfies

$$\int_0^\infty H_\Lambda(B^*(s)z^*(s))ds \geq \epsilon.$$

Thus every such solution $z^*(t)$ also satisfies the following inequalities.

$$\begin{aligned}
\int_0^\infty H_\Lambda(B^*(s)z^*(s))ds &\geq \int_0^\infty H_{B(\delta)}(B^*(s)z^*(s))ds \\
&\geq \int_0^\infty \text{Sup}\ \{\langle\alpha, B^*(s)z^*(s)\rangle \mid ||\alpha|| < \delta\}ds \\
&\geq \frac{\delta}{R}\int_0^\infty \text{Sup}\ \{\langle\alpha, B^*(s)z^*(s)\rangle \mid ||\alpha|| < R\}ds \\
&\geq \frac{\delta}{R}\int_0^\infty H_{B(R)}(B^*(s)z^*(s))ds \\
&\geq \frac{\delta\epsilon}{R}.
\end{aligned}$$

Another application of Lemma 4.2 completes the proof. □

Now we prove the main theorem of this section.

4.4 Theorem. *Consider the linear control process* $(2.1.\omega)$ *with constraint set* Λ. *Suppose the following conditions are satisfied*

(1) *The flow* $(\Omega, \{T_t\}_{t\in\mathbf{R}})$ *is minimal.*
(2) *The set* Λ *is compact and* $0 \in \Lambda$.
(3) *For some* $\omega_0 \in \Omega$ *the process* $(2.1.\omega_0)$ *is locally null controllable.*

Then the process is uniformly locally null controllable.

Proof. For convenience we set,

$$f_\omega^p(t) = B^*(T_t(\omega))X_A^*(\omega,t)^{-1}p, \qquad p \in \mathbf{R}^n,\ \omega \in \Omega.$$

Now, Lemma 4.2 and assumption (3) implies the existence of an $\epsilon > 0$ such that

$$\int_0^\infty H_\Lambda(f_{\omega_0}^p(s))ds \geq 8\epsilon \qquad \text{for all } p \in \mathbf{R}^n \text{ with } ||p|| = 1.$$

Since the map $p \to H_\Lambda(f_{\omega_0}^p(s))$ is continuous and non-negative for each s, compactness of the unit sphere in $\mathbf{R}^n$ implies that there exists $J > 0$ such that

$$\int_0^J H_\Lambda(f_{\omega_0}^p(s))ds \geq 4\epsilon \qquad \text{for all } p \in \mathbf{R}^n \text{ with } ||p|| = 1.$$

Let d be a metric on Ω generating its topology. Using the continuity of the map $\omega \to \int_0^J H_\Lambda(f_\omega^p(s))ds$, we can find $\delta > 0$ such that, if $d(\omega,\omega_0) \leq \delta$ then the values of this map at ω and ω_0 are within ϵ for all p with $||p|| = 1$.

Now minimality of the flow implies that the point ω_0 is uniformly recurrent. Therefore, corresponding to the number δ just defined, there exists $L > 0$ and a sequence $t_n \to -\infty$ with the following properties:

(1) $t_n < t_{n-1}$,
(2) $|t_n - t_{n-1}| < L$,
(3) $T_{t_n}(\omega_0)$ lies in the δ-ball centered at ω_0 for all $n \in \mathbf{N}$.

These conditions along with the choice of δ implies that

$$\int_0^J H_\Lambda(f_{T_{t_n}(\omega_0)}^p(s))ds \geq 2\epsilon \qquad \text{for all } p \in \mathbf{R}^n \text{ with } ||p|| = 1 \text{ and } n \in \mathbf{N}. \tag{4.1}$$

Next, let

$$M = \sup\left\{max\ \{||X_A^*(\omega,r)^{-1}||, 1\} \;\middle|\; r \in [-L,0],\ d(\omega,\omega_0)) \leq \delta\right\}.$$

We claim that

$$(4.2)\qquad \int_0^{J+L} H_\Lambda(f^p_{T_t(\omega_0)}(s))ds \geq \frac{2\epsilon}{M} \qquad \text{for all } p \in \mathbf{R}^n \text{ with } ||p|| = 1 \text{ and } t < 0.$$

To see this, note that for each $\omega \in \Omega$ and $r < 0$ we have

$$\begin{aligned} H_\Lambda(f^p_{T_r(\omega)}(s)) &= H_\Lambda\big(B^*(T_s(T_r(\omega)))X^*_A(T_r(\omega), s)^{-1}p\big) \\ &= H_\Lambda\big(B^*(T_{s+r}(\omega))X^*_A(\omega, r+s)^{-1}X^*_A(\omega, r)p\big). \end{aligned}$$

Here we have used the cocycle identity for $X_A^{*\,-1}$. Thus, letting $\tilde{p} = X^*(\omega, r)p$, we get

$$\begin{aligned} &\inf\{\int_0^{J+L} H_\Lambda(f^p_{T_r(\omega)}(s))ds \mid ||p|| = 1\} \\ &\qquad = \inf\{\int_0^{J+L} H_\Lambda(f^{\tilde{p}}_\omega(r+s))ds \mid ||p|| = 1\} \\ &\qquad = \inf\{\int_r^{r+J+L} H_\Lambda(f^{\tilde{p}}_\omega(s))ds \mid ||p|| = 1\} \\ &\qquad \geq \inf\{\int_0^{r+J+L} H_\Lambda(f^{\tilde{p}}_\omega(s))ds \mid ||p|| = 1\} \quad (\text{since } r < 0) \\ &\qquad \geq \inf\{||X^*_A(\omega, r)p|| \int_0^{r+J+L} H_\Lambda(f^q_\omega(s))ds \mid ||p|| = 1\}, \end{aligned}$$

where $q = p/||\tilde{p}||$. Now, give any $t < 0$, write $t = t_n + r$ for some n where $r \in [-L, 0]$. Applying the above inequality with $\omega = T_{t_n}(\omega_0)$, we obtain the following estimates:

$$\begin{aligned} &\inf\{\int_0^{J+L} H_\Lambda(f^p_{T_t(\omega_0)}(s))ds \mid ||p|| = 1\} \\ &\qquad = \inf\{||X^*_A(T_{t_n}(\omega_0), r)p|| \int_0^{r+J+L} H_\Lambda(f^q_{T_{t_n}(\omega_0)}(s))ds \mid ||p|| = 1\} \\ &\qquad \geq 2\epsilon \inf\big\{||X^*_A(T_{t_n}(\omega_0), r)p|| \mid ||p|| = 1\big\} \qquad (\text{by } (4.1)), \\ &\qquad \geq 2\epsilon \inf\big\{||X^*_A(T_{t_n}(\omega_0), r)p|| \mid ||p|| = 1,\ r \in [-L, 0]\big\}. \end{aligned}$$

Since $||p|| \leq ||N^{-1}||\ ||Np||$ for any nonsingular matrix N, we finally obtain,

$$\inf\{\int_0^{J+L} H_\Lambda(f^p_{T_t(\omega_0)}(s))ds \mid ||p|| = 1\} \geq \frac{2\epsilon}{M}.$$

Thus, inequality (4.2) is proved.

Now, minimality of $(\Omega, \{T_t\}_{t\in\mathbf{R}})$ implies that the set $\{T_t(\omega_0) \mid t < 0\}$ is dense in Ω [E]. Again using continuity of the map $\omega \to \int_0^{J+L} H_\Lambda(f^p_\omega(s))ds$, we obtain

$$(4.3)\qquad \int_0^{J+L} H_\Lambda(f^p_\omega(s))ds \geq \frac{2\epsilon}{M}, \qquad \text{for all } \omega \in \Omega \text{ and } ||p|| = 1.$$

Now define the set $V = \{x \in \mathbf{R}^n \mid ||x|| < \frac{\epsilon}{M}\}$. We shall show that $V \subseteq D(\omega, T)$ for all $\omega \in \Omega$, where T is independent of ω. This will complete the proof of Theorem 4.4. We do so by essentially repeating the steps in the proof of Lemma 4.2(b) (see [BS2]).

Note that Lemma 4.2(a) reduces our proof to showing there exists a $T > 0$ such that, for each $x \in V$ and $\omega \in \Omega$ the following inequality holds,

$$\langle x, p\rangle + \int_0^T H_\Lambda(f_\omega^p(s))ds \geq 0, \qquad \text{for all } p \in \mathbf{R}^n \text{ with } ||p|| = 1.$$

If this inequality is false, then there exist $x_0 \in V$ and sequences $p_n \in \{p \in \mathbf{R}^n \mid ||p|| = 1\}$, $\omega_n \in \Omega$ and $\hat{T}_n \to \infty$ such that

$$\int_0^{\hat{T}_n} H_\Lambda(f_{\omega_n}^{p_n}(s))ds < -\langle x_0, p_n\rangle \leq ||x_0|| < \frac{\epsilon}{M}, \qquad \text{for all } n \in \mathbf{N}.$$

By compactness of Ω and the unit sphere in $\mathbf{R}^n$, without loss of generality assume that $p_n \to p$ and $\omega_n \to \omega$. Thus,

$$\lim_{n\to\infty} H_\Lambda(f_{\omega_n}^{p_n}(s))\, 1_{[0,\hat{T}_n]}(s) = H_\Lambda(f_\omega^p(s)),$$

where $1_{[0,\hat{T}_n]}$ is the characteristic function of the set $[0, \hat{T}_n]$ and the limit is a pointwise limit on $\mathbf{R}$. We thus have;

$$\begin{aligned}\int_0^\infty H_\Lambda(f_\omega^p(s))ds &\leq \liminf_{n\to\infty} \int_0^{\hat{T}_n} H_\Lambda(f_{\omega_n}^{p_n}(s))ds \\ &\leq \limsup_{n\to\infty} \int_0^{\hat{T}_n} H_\Lambda(f_{\omega_n}^{p_n}(s))ds \\ &\leq ||x_0|| \\ &< \frac{\epsilon}{M},\end{aligned}$$

where the first inequality uses Fatou's lemma. However this violates inequality (4.3). The proof of Theorem 4.4 is now complete. □

One may ask whether in the above theorem, the assumption of minimality on the base flow $(\Omega, \{T_t\}_{t\in\mathbf{R}})$ can be weakened? The following proposition answers this question if the controls are not constrained.

4.5 Proposition. *Suppose that for each minimal subset M of Ω, there exists at least one $\omega_0 \in M$ such that the process $(2.1.\omega_0)$ is locally null controllable. Then the process $(2.1.\omega)$ is uniformly locally null controllable.*

Proof. In the absence of constraints, controllability is determined by the non-singularity of the following **controllability matrix**;

$$W_T(\omega) = \int_0^T X_A(\omega, t)^{-1} B(T_t(\omega)) B^*(T_t(\omega)) X_A^*(\omega, t)^{-1} dt.$$

We claim that there exists constants $\alpha > 0$ and $T > 0$ such that

$$0 < \alpha I < W_T(\omega), \qquad \text{for all } \omega \in \Omega.$$

Suppose for contradiction, there exist sequences $\omega_j \in \Omega$, $T_j \to \infty$, $\alpha_j \to 0$ and $x_j \in \mathbf{R}^n$ such that $||x_j|| = 1$ and $||W_{T_j}(\omega_j)x_j|| \leq \alpha_j$. Then by passing to a subsequence if necessary, we can find a minimal set M of Ω, $\overline{\omega} \in M$ and $\overline{x} \in \mathbf{R}^n$, $||x|| = 1$, such that $\omega_j \to \overline{\omega}$ and $x_j \to \overline{x}$. Notice that

$$W_T(\overline{\omega})\overline{x} = 0 \quad \text{for all } T > 0. \tag{4.4}$$

Now our assumption allows us to apply Theorem 4.4 where the base flow is $(M, \{T_t\}_{t\in\mathbf{R}})$. In particular this implies that the process (2.1.$\overline{\omega}$) is null controllable. Thus for some $T > 0$ the controllability matrix $W_T(\overline{\omega})$ is nonsingular. This contradicts equation (4.4). Thus our claim is proved. The proof follows immediately from this claim. □

Proposition 4.5 will be used below in our discussion of the feedback stabilization problem. Finally we remark that Proposition 4.5 remains true in the presence of constraints. This is proved in a recent paper by F. Colonius and R. Johnson [CJ]. In this paper it is also shown how the continuity hypothesis on B can be weakened in Theorem 4.4 and Proposition 4.5; one uses the theory of "measurable selections".

§5. Dependence of Controllability on The Dynamics of The Flow.

Consider the random linear control process given by (2.1.ω). Keeping the functions A and B fixed, one can vary the flow on Ω and ask; how sensitive is the set of vectors that can be steered to zero, to the dynamics of the flow on Ω. We shall now demonstrate that this dependence can be strikingly delicate. We shall let Ω to be the two torus $\mathbf{S}^1 \times \mathbf{S}^1$ and consider the family of rotation flows $\{T_t^\alpha\}_{t\in\mathbf{R}}$ on Ω parametrized by their irrational winding number α. These flows are given by

$$T_t^\alpha(x, y) = (x + t, y + \alpha t) \ mod\ 1.$$

Here we are thinking of $\mathbf{S}^1$ as the interval $[0, 1]$ whose endpoints are identified. We shall construct smooth functions A and B on Ω such that the controllability properties of the control process

$$x' = A(T_t^\alpha \omega)x + B(T_t^\alpha \omega)u, \qquad \omega \in \Omega, \tag{5.1.ω.α}$$

depend on the continued fraction properties of the irrational number α. More precisely we have the following theorem [N3].

5.1 Theorem. *Let β and γ be given irrationals such that β is Liouville and γ is diophantine. Then there exists a linear control system of the form (5.1.ω.α) with C^∞ coefficient functions A and B such that the following holds.*

(1) *If $\alpha = \beta$, then the system is globally null controllable for all $\omega \in \Omega$. In fact, given any $r > 0$, the system is globally null controllable by a bounded control with constraint $|u| < r$.*
(2) *If $\alpha = \gamma$ then the system is not globally null controllable for any $\omega \in \Omega$.*

We recall the following definition.

5.2 Definition. *(a) An irrational β is Liouville if there is a sequence $\frac{p_n}{q_n}$ of rationals with $q_n \to \infty$ and a positive constant K such that*

$$|\beta - \frac{p_n}{q_n}| \leq \frac{K}{q_n^n}, \quad \text{for all } n \in \mathbf{N}.$$

(b) An irrational γ is diophantine if there exist positive constants L and η, $(0 < \eta < 1)$ such that the following inequality holds for all irreducible fractions $\frac{p}{q}$;

$$|\gamma - \frac{p}{q}| > \frac{L}{q^{2+\eta}}.$$

Now we sketch the idea behind the proof. For simplicity we shall consider only two-dimensional processes. First let us consider the process without any constraints. Notice that the obstruction to controllability is precisely those solutions $x_\omega(t)$ of the linear system adjoint to the one corresponding to $(5.1.\omega.\alpha)$, for which the following holds:

$$x_\omega(t) \in \text{Ker } B^T(T_t^\alpha \omega), \quad \text{for all } t \geq 0.$$

This follows from the proof of the standard characterization of controllability in terms of the non-singularity of the controllability matrix [LM]. Thus $(5.1.\omega.\alpha)$ is not controllable if and only if there exists a closed set $M \subseteq \mathbf{P} \times \Omega$, invariant under the skew- product flow generated by the adjoint linear system such that

$$M \subseteq \{([v], \omega) \in \mathbf{P} \times \Omega \mid v \in \text{Ker } B^T(\omega)\}.$$

Thus, if we show that this adjoint skew-product flow is minimal, then that implies null controllability, (of course, provided $B(\omega)$ is not identically zero). Existence of a C^∞ function A on Ω which generates a minimal "adjoint skew-product flow" provided the winding number β of the base flow is a Liouville number, is the main result of [N3]. Furthermore, this map A can be taken to be of the form

$$\begin{pmatrix} 0 & f(\omega) \\ -f(\omega) & 0 \end{pmatrix},$$

for a suitable C^∞ map $f : \Omega \to \mathbf{R}$ with $\int_\Omega f(\omega) d\mu = 0$, where μ is the usual Lebesgue measure on Ω. This construction is rather technical and is motivated by some other constructions in Ergodic Theory. Granting this construction, we indicate how Theorem 5.1 is proved.

We recall a classical result based on the "small divisor argument". The result says that if the winding number γ of the flow on Ω is diophantine, smoothness of f guarantees existence of a smooth function $g : \Omega \to \mathbf{R}$ such that

$$f(\omega) = D_\gamma g(\omega) \equiv \lim_{h \to 0} \frac{g(T_h^\gamma(\omega)) - g(\omega)}{h}.$$

This suggests that if we take our control process to be

$$x' = \begin{pmatrix} 0 & f(T_t^\alpha(\omega)) \\ -f(T_t^\alpha(\omega)) & 0 \end{pmatrix} x + \begin{pmatrix} \cos g(T_t^\alpha(\omega)) & \sin g(T_t^\alpha(\omega)) \\ -\sin g(T_t^\alpha(\omega)) & \cos g(T_t^\alpha(\omega)) \end{pmatrix} Bu,$$

where B is a constant, rank one matrix, then for $\alpha = \beta$ the process is globally null controllable. However for $\alpha = \gamma$, the following change of variables

$$y = \begin{pmatrix} \cos g(T_t^\alpha(\omega)) & \sin g(T_t^\alpha(\omega)) \\ -\sin g(T_t^\alpha(\omega)) & \cos g(T_t^\alpha(\omega)) \end{pmatrix} x$$

transforms the process $(5.1.\omega.\alpha)$ to the system

$$y' = Bu$$

and this system is not null controllable because $\text{rank}(B) = 1$.

Now let us consider the magnitude constraints in part (1) of the theorem. Notice that the constraint set $\Lambda = B_r(0)$-the ball of radius r centered at the origin, satisfies the conditions of Lemma 4.2. Hence if the process $(5.1.\omega.\beta)$ is not null controllable, by Lemma 4.2(b) we can find a sequence $x_n \in \mathbf{R}^2$ with $||x_n|| = 1$ such that

$$\int_0^\infty ||B^*(T_t^\beta(\omega))X_A^*(T_t^\beta(\omega))x_n||dt \leq \frac{1}{n}, \quad n \in \mathbf{N}.$$

Thus, fixing any large $T > 0$ and using the usual compactness and continuity argument and passing to a convergent subsequence we get a point $x_0 \in \mathbf{R}^2$ with $||x_0|| = 1$ such that

$$\int_0^T ||B^*(T_t^\beta(\omega))X_A^*(T_t^\beta(\omega))x_0||dt = 0.$$

Since $T > 0$ is arbitrary, this contradicts controllability of $(5.1.\omega.\beta)$. □

We note that the dichotomy spectrum of our linear system is the singleton set $\{0\}$, irrespective of the nature of the winding number α. The results proved in Section 6 point out that given local null controllability, global null controllability depends only on the dichotomy spectrum of the corresponding linear system. Thus in the example constructed above, the essential difference between control systems for different values of the irrational α lies in their different local null controllability behaviour.

§6. Global Null Controllability.

Now we discuss the relationship between local and global null controllability. This is well understood when A and B are constant matrices and the constraint set Λ is compact, convex and contains zero. In this case it is known that global null controllability is equivalent to local null controllability together with the condition that the real parts of the eigenvalues of A are all non-positive. Using local null controllability, it is easy to steer generalized eigenvectors of A corresponding to eigenvalues with negative real part to zero. However, steering generalized eigenvectors with purely imaginary eigenvalues to zero is not quite as simple. In this case, for constant coefficient systems, one can use arguments based on the Jordan normal form or on the convexity of the reachable set (see [LM]).

The above discussion points out two general principles which carry over to the study of non-autonomous control processes. First, once one is given local null controllability, the global null controllability depends only on the qualitative behaviour of the solution of the associated linear system and has nothing to do with matrix B. Second, it may in general be difficult to steer vectors with "zero exponents" to zero because they may exhibit a sort of "transient" behaviour. Thus the additional condition one needs to impose is that these solutions of $(3.1.\omega)$ should be recurrent in some sense. This in turn suggests that ergodic theory can play a role in studying our controllability question.

In the language of qualitative theory of random linear systems developed earlier, the additional condition one needs to impose is the following: the zero exponent is the right endpoint of the spectral sub-interval to which it belongs. Translating this condition in to the language of ergodic theory and then, using an ergodic-theoretic recurrence result, we shall see that norm of a vector with zero exponent returns close to its initial value after a sufficiently long time. This fact will enable us to steer such vectors closer to zero than before, and finally to zero in finitely many steps (due to local null controllability). Basically this is the idea involved in the proof of the next Theorem.

6.1 Theorem. *Consider the control process* $(2.1.\omega)$ *with constraint set* $\Lambda \subseteq \mathbf{R}^n$. *Suppose that,*

(1) Λ *is compact and* $0 \in \Lambda$,
(2) *the family of control processes* $(2.1.\omega)$ *is uniformly locally null controllable and*
(3) *the dichotomy spectrum* Σ *of the corresponding linear system* $(3.1.\omega)$ *is contained in* $(-\infty, 0]$.

Let μ *be an invariant measure on* Ω. *Then the process* $(2.1.\omega)$ *is globally null controllable for* μ *almost all* $\omega \in \Omega$.

We remark that if all solutions to the equations $(3.1.\omega)$ are bounded as t varies over $\mathbf{R}$, then it can be shown that the system $(2.1.\omega)$ is globally null controllable **for all** $\omega \in \Omega$.

We shall discuss the proof in some detail along with the proof of another theorem to come (Theorem 6.3). But first, we have the following partial converse to Theorem 6.1.

6.2 Theorem. *Suppose the flow* $(\Omega, \{T_t\}_{t\in\mathbf{R}})$ *is minimal and supports exactly one invariant measure* μ, *which is then necessarily ergodic.. For example, this is the case when the flow is the translation flow on the hull of Bohr almost periodic functions. Suppose that*

$$\mu\{\omega \in \Omega \mid \text{the process } (2.1.\omega) \text{ is globally null controllable}\} > 0.$$

Then the dichotomy spectrum Σ *of* $(3.1.\omega)$ *is contained in* $(-\infty, 0]$.

Proof. Assume that the spectrum Σ intersects the positive real axis. Then the spectrum Σ^* of the adjoint linear system meets the negative real axis. Let $b < 0$ be the left end point of one of the intervals in Σ^*. Then by [JPS] there is an ergodic probability measure ν on $\mathbf{P} \times \Omega$ such that

$$\lim_{t\to\infty} \frac{1}{t}\, ln\, ||X_A^*(\omega,t)^{-1}v|| = b < 0, \quad \nu \text{ a.e. } (v,\omega) \in \mathbf{P} \times \Omega.$$

Here we confuse $v \neq 0$ with the line in $\mathbf{R}^n$ on which it lies. Since ν projects on to μ we have

$$\mu\{\omega \in \Omega \mid \lim_{t\to\infty} \frac{1}{t}\, ln\, ||X_A^*(\omega,t)^{-1}v|| < 0 \text{ for some } v \in \mathbf{R}^n - \{0\}\} = 1.$$

Now, Proposition 2.1 of [Pa] shows that if (2.1.ω) is globally null controllable at ω, then $\lim_{t\to\infty} \frac{1}{t}\, ln\, ||X_A^*(\omega,t)^{-1}v|| \geq 0$ for all non zero $v \in \mathbf{R}^n$. Thus

$$\mu\{\omega \in \Omega \mid \lim_{t\to\infty} \frac{1}{t}\, ln\, ||X_A^*(\omega,t)^{-1}v|| \geq 0 \text{ for all } v \in \mathbf{R}^n - \{0\}\} > 0.$$

This contradicts the previous paragraph and completes the proof. □

Now we discuss the most striking result. Suppose that (2.1.ω) is locally null controllable. Futher assume that the linear system (3.1.ω) admits solutions with positive exponential growth; i.e. has at least one positive Lyapunov exponent. The question is can (2.1.ω) be globally null controllable?

One's first response might well be "never"; global null controllability can not hold because of the presence of positive Lyapunov exponents. If $A(t) \equiv A$ is a constant function, then the presence of a positive Lyapunov exponent means that some eigenvalue of A has positive real part. In this case, it is well known that (2.1) can not be globally null controllable. Similarly if $A(t)$ is periodic, then positivity of its Floquet exponent prohibits (2.1) from being globally null controllable. However, we will see that in the case of non-periodic recurrence of the function $A(t)$, it is rather (topologically) often the case that (2.1.ω) is globally null controllable even when (3.1.ω) admits positive Lyapunov exponents. More precisely, if (3.1.ω) satifies the irreducibility condition (Definition 3.7) and if the dichotomy spectrum is an interval containing 0, then "topologically most" (i.e. residually many) processes in the family (2.1.ω) are globally null controllable.

In some sense there is a similarity between Theorem 6.1 and Theorem 6.3. below. Both theorems conclude global null controllability of (2.1.ω) for a "large" set of ω's, where "large" is understood in the measure theoretic and topological sense in the former and the later theorem respectively. Furthermore, the above partial converse to Theorem 6.1 points out that if a system satisfies the hypotheses of Theorem 6.3 and if $d > 0$ (see the hypothesis of Theorem 6.3), then this topologically large set of ω's will have measure zero. There is another common point between the proofs of these two theorems. Both proofs are based on recurrence results, in the former case a measure theoretic one and in the later case a topological one.

Recall that, given local null controllability, our interest is in conditions assuring global null controllability, especially when equations (3.1.ω) admit a positive Lyapunov exponent. Suppose that the dichotomy spectrum of equations (3.1.ω) is

$$\Sigma = [c_1, d_1] \cup \cdots \cup [c_p, d_p],$$

where $c_1 \le d_1 < c_2 \le \cdots < c_p \le d_p$. Furthermore, suppose that $0 \in [c_r, d_r]$, $1 \le r \le p$. Let $W_1, \cdots, W_p$ be the corresponding subbundles of $\mathbf{R}^n \times \Omega$. If $(v_0, \omega) \in W_i$ and $i < r$, then v_0 can be easily steered to zero, by first defining the control u to be zero untill v_0 enters a sufficiently small neighbourhood of zero and then defining it appropriately using local null controllability. On the other hand if $i > r$, then we leave as an exercise to show that, if $||v_0||$ is large enough, then v_0 cannot be steed to zero.

It is natural, then, to restrict attention to the case when Σ is a single interval $[c, d]$, i.e. to restrict attention to the subbundle W_r. The reader should be warned that there are problems with this process. Especially the bundle $W_r \subseteq \mathbf{R}^n \times \Omega$ need not be topologically trivial (for example see [EJ], [P1]), and thus it not clear how to "restrict" equations (3.1.ω) to W_r. However, one can trivialize W_r with a recurrent change of variables $x = C(t)w$ (see [P1]), and this is usually sufficient to treat solutions in W_r as if they arose from a family (3.1.ω) with $\Sigma = [c_r, d_r]$.

Thus, to discuss the non-trivial part of our results it is enough to suppose that $\Sigma = [c, d]$ is an interval containing zero. We are particularly interested in global null controllability of (2.1.ω) when $c \le 0 < d$. Notice that in this case, (3.1.ω) admits a positive Lyapunov exponent $\beta = d$. The following is our surprising "topological conterpart" of Theorem 6.1.

Theorem 6.3. *Consider the control processes* (2.1.ω). *Suppose that*

(1) *the process is locally null controllable (for some ω),*
(2) *the dichotomy spectrum $\Sigma = [c, d]$ is a single interval and*
(3) *there exists $\beta \le 0$ such that $\beta \in \cap\{J_i \mid 1 \le i \le n\}$, (in particular* (3.1.$\omega$) *is irreducible).*

Then there is a residual set $\Omega_1 \subseteq \Omega$ such that if $\omega \in \Omega_1$ then (2.1.ω) *is globally null controllable.*

Sketch of Proofs of Theorems 6.1 and 6.3. The proof of both, Theorem 6.1 and Theorem 6.3 uses the technique of "recurrent triangularization" discussed before in Section 3. Recall that $M \subseteq Y = O(n) \times \Omega$ is a minimal set and $\pi : M \to \Omega$ is the projection $\pi(g, \omega) = \omega$. The orthogonal change of variables $x = F(\hat{T}_t(m))v$ which transforms (3.1.ω) (lifted to M via π) to (3.5.m) takes the control process (2.1.ω) (lifted to M via π) to

$$v' = r(\hat{T}_t(m))v + F^{-1}(\hat{T}_t(m))B \circ \pi(\hat{T}_t(m))u. \tag{6.1.m}$$

Clearly (2.1.ω) is locally (respectively globally) null controllable if and only if (6.1.m) is so.

Next write v in the component form, $v = (v_1, \cdots, v_n)^T$. Observe that the first component v_1 satisfies

$$v_1' = r_{1,1}(\hat{T}_t(m))v_1, \quad m \in M. \tag{6.2.m}$$

Now we give the arguments involved in the proofs of both theorems indicating the type of recurrence result used.

Proof of Theorem 6.3 We shall need the following topological recurrence result [J1].

Proposition 6.4. *Let $(X, \{T_t\}_{t\in\mathbf{R}})$ be a minimal flow and $R : X \to \mathbf{R}$ be a continuous function such that $\int_X R d\mu \leq 0$ for some flow invariant, Borel probability measure μ on X. Then exactly one of the following two possibilities hold.*

(1) *There exists a residual subset $X_1 \subseteq X$ such that*

$$\liminf_{t\to\infty} \int_0^t R(T_s(x))ds = -\infty, \quad \textit{for all } x \in X_1.$$

(2) *There exists a continuous function $\tilde{R} : X \to \mathbf{R}$ such that*

$$\tilde{R}(T_t(x)) - \tilde{R}(x) = \int_0^t R(T_s(x))ds, \quad \textit{for all } x \in X.$$

We apply this lemma to $(M, \{\hat{T}_t\}_{t\in\mathbf{R}})$ with $R = r_{1,1}$. Our irreducibility asssumption and Proposition 3.8 implies that the hypotheses of Proposition 6.4 is satisfied. Suppose that the first alternative in Proposition 6.4 holds. This implies that for residually many m's in M, the solution $v_1^m(t) \equiv v_1(t)$ tends to the origin with zero control along some sequence of times tending to infinity. Hence once $v_1(t)$ gets sufficiently close to the origin, using local null controllability it can be steered to the origin. This argument shows that for residually many m's in M, v can be m-steered to a vector whose first component is zero.

Now suppose the second alternative in Proposition 6.4 holds. Let $\delta > 0$ be so small that any point in the ball of radius δ cantered at the origin can be steered to the origin. Now using minimality of $(M, \{\hat{T}_t\}_{t\in\mathbf{R}})$ we can find a large enough time $T_1 > 0$ such that

$$|v_1(T_1) - v_1(0)| < \frac{\delta}{2}.$$

Using local null controllability we can steer $v = (v_1, \cdots, v_n)^T$ to a vector $v(T_1) = (\tilde{v}_1, \cdots, \tilde{v}_n)^T$ satisfying

$$|\tilde{v_1}| < (1 - \frac{\delta}{2})|v_1|.$$

Repeating this construction at most $[\frac{2|v_1|}{\delta}]$ times, as before, we can steer v to a vector whose first component zero.

We can now apply above argument to the second component of the vector. Using the fact that $\{v \mid v_1 = 0\}$ is invariant under the solutions of the linear system corresponding to the process (6.1.m) and using the analogue of the previous argument, v can be m-steered to a vector with first two components zero for residually many $m \in M$.

Repeating n times the above arguments, we obtain a control which steers v to zero in finite time. Thus there is a residual set $M_1 \subseteq M$ of points such that (6.1.m) is globally null controllable. Now, since M is minimal, the projection $\Omega_1 \equiv \pi(M_1)$ is residual. This completes the proof of Theorem 6.3.

Proof of Theorem 6.1 This proof resembles the proof in the case of the second alternative above. We need the following ergodic theoretic recurrence result; see [Sc] for a proof.

Proposition 6.5. *Let $(X, \{T_t\}_{t\in\mathbf{R}}, \mu)$ be a flow on a compact metric space X with invariant, ergodic, Borel probability measure μ. Let $R : X \to \mathbf{R}$ be a μ integrable function such that $\int_X R d\mu \leq 0$. Let*

$$\tilde{X} = \{x \in X \mid \text{given } \epsilon > 0 \text{ and } N > 0, \int_0^t R(T_s(x))ds \leq \epsilon \text{ for some } t > N\}.$$

Then $\mu(\tilde{X}) = 1$.

We apply this proposition to $(M, \{\hat{T}_t\}_{t\in\mathbf{R}})$ with $R = r_{1,1}$. The consequences of this are same as those of the "second alternative" in the proof of Theorem 6.3. The only difference is that in this case instead of a residual subset we get a subset $M_1 \subseteq M$ of **full measure**. The rest of the proof remains essentially the same as in the case of the second alternative above. □

Now we consider some consequences of Theorem 6.3.

Corollary 6.6. *Consider the control process (2.1.ω). Suppose that*

(1) *the flow $(\Omega, \{T_t\}_{t\in\mathbf{R}})$ admits a unique invariant, Borel probability measure,*
(2) *let $n = 2$ and that $0 \in \Sigma = [c, d]$ and*
(3) *the process (2.1.ω) is locally null controllable.*

Then there is a residual subset $\Omega_1 \subseteq \Omega$ such that (2.1.ω) is globally null controllable for all $\omega \in \Omega_1$.

Proof. This corollary follows from Proposition 3.11(a) and Theorem 6.3. □

We also have a specific result when $n = 3$.

Corollary 6.7. *Consider the control process (2.1.ω). Suppose that*

(1) *the flow $(\Omega, \{T_t\}_{t\in\mathbf{R}})$ admits a unique invariant, Borel probability measure μ,*
(2) *let $n = 3$ and that there is only one spectral interval in the dichotomy spectrum and*
(3) *the process (2.1.ω) is locally null controllable.*

Then there is a residual subset $\Omega_1 \subseteq \Omega$ such that (2.1.ω) is globally null controllable for all $\omega \in \Omega_1$.

Proof. This follows from Proposition 3.11(b) and Theorem 6.3. □

We state a final result, this time in arbitrary dimension n. Again, suppose that $\Sigma = [c, d]$ and that $(\Omega, \{T_t\}_{t\in\mathbf{R}})$ admits a unique invariant probability measure μ. Suppose that there are n **distinct** Lyapunov exponents $\beta_1 < \beta_2 \cdots < \beta_n$. (Millionschikov [M2] shows that this condition is generic for almost periodic systems.) Using Proposition 3.9, one can show that each interval J_i must contain β_r, where $1 \leq r \leq n - 1$. Hence, if $\beta_{n-1} \leq 0$, the proof of Theorem 6.3 applies and one concludes that (2.1.ω) is globally null controllable for a residual subset of ω's in Ω. Observe that irreducibility is not required for this argument.

Finally, as mentioned before, we observe that if $(\Omega, \{T_t\}_{t\in\mathbf{R}})$ is uniquely ergodic, $\Sigma = [c,d]$ and if $d > 0$, then the residual subset $\Omega_1 \subseteq \Omega$ obtained above satisfies $\mu(\Omega_1) = 0$. This follows from Theorem 6.2.

§7. The Feedback Stabilization Problem for Random Linear Systems.

The purpose of this section is to consider the random feedback stabilization problem. We give a solution based on a random version of the classical linear regulator problem, and on the less classical ingredients of **exponential dichotomy** and **rotation number** for linear Hamiltonian systems [JN3]. In this approach we avoid a direct treatment of the Riccati equation which has been a canonical tool for studying the linear regulator problem. Instead we appeal to the now well-developed theory of random linear Hamiltonian systems. We get very good smoothness and robustness results for the stabilizing feedback matrix K. These results are obtained as direct applications of standard theorems in the theory of exponential dichotomy [Cp,P2,P3,Y].

We begin by formulating the feedback stabilization problem in the language of this article. Consider the equation (2.1)

$$x' = a(t)x + b(t)u \qquad x \in \mathbf{R}^n,\ u \in \mathbf{R}^m,$$

where the coefficients $a(\cdot)$ and $b(\cdot)$ satisfy the conditions set out in Section 2. We seek a control of the form $u = k(t)x$, where $k(\cdot)$ is an $m \times n$ matrix valued function of t, such that $x = 0$ is an exponentially stable solution of the equation

$$x' = [a(t) + b(t)k(t)]x. \tag{7.1}$$

We introduce the framework of random linear control processes discussed in Section 2. Instead of (2.1), we consider the family of control processes (2.1.ω)

$$x' = A(T_t(\omega))x + B(T_t(\omega))u \qquad (\omega \in \Omega),$$

where $(\Omega, \{T_t\}_{t\in\mathbf{R}})$ is a flow and A, B satisfy conditions set out in Section 2. We pose the problem of finding a **continuous** mapping $K : \Omega \to M(m,n)$ with the property that, for each $\omega \in \Omega$, the linear feedback control $u(t,x) = K(T_t(\omega))x$ stabilizes (2.1.ω). That is, we require that $x = 0$ be a uniformly exponentially stable solution of the equation

$$x' = [A(T_t(\omega)) + B(T_t(\omega))K(T_t(\omega))]x \tag{7.1.ω}$$

for every $\omega \in \Omega$.

The condition that K be a continuous function of ω is clearly a natural one, and of course is much stronger than requiring the existence of a feedback control $u(t) = k(\omega,t)x$ for each $\omega \in \Omega$. The continuity in ω of the function $K(\omega)$ is what we refer to as "conservation of recurrence".

The feedback stabilization problem for a single control process (2.1) can be treated by embedding it in a random family, as described in Section 2. Any recurrence properties of the coefficients $a(\cdot)$, $b(\cdot)$ will be reflected in the structure of the compact metric space Ω, and this structure in turn will be reflected by a **continuous** feedback matrix K. This is the motivation for the term "conservation of recurrence".

To illustrate this property more concretely, and also to illustrate our smoothness and robustness results, we consider the following situation. Let Ω be a C^∞ compact manifold, and let X be a C^∞ vector field on Ω. Let $\{T_t\}_{t\in\mathbf{R}}$ be the one-parameter group of diffeomorphisms of Ω generated by X. Suppose that $(\Omega, \{T_t\}_{t\in\mathbf{R}})$ is "chaotic" in one of the accepted senses of the term, e.g., it exhibits sensitive dependence on initial conditions, has positive entropy, etc.

Next let $A : \Omega \to M(n,n)$ and $B : \Omega \to M(n,m)$ be C^∞ functions, and consider the corresponding family of control processes (2.1.ω). It will be shown later on that, subject to a controllability condition on (2.1.ω), there exists a continuous (in fact C^∞) function $K : \Omega \to M(m,n)$ such that, for each $\omega \in \Omega$, $u(t) = K(T_t(\omega))x$ stabilizes (2.1.ω). The continuity in ω of $K(\cdot)$ means that $k(t) = K(T_t(\omega))$ is "no more chaotic" than $a(t) = A(T_t(\omega))$, $b(t) = B(T_t(\omega))$ for each $\omega \in \Omega$: this is an example of conservation of recurrence.

Smoothness is illustrated by the fact that, if A and B depend in a C^r manner on a parameter $\lambda \in \mathbf{R}^p$, then K also depends in a C^r manner on λ, $(0 \le r \le \infty)$. Moreover if $A(\cdot)$ and $B(\cdot)$ are C^r in ω, then $K(\cdot)$ may be chosen to be C^r in ω. These statements follow from basic results in the theory of exponential dichotomy ([Cp,CY,P2,P3,Y]). Similarly, the theory of exponential dichotomy allows us to conclude that, if A and B are perturbed in, say, $C^0(\Omega)$, then K varies in a Lipschitz manner on these perturbations ([Cp]).

Let us now turn to the construction of the feedback matrix K. We will use the linear regulator problem to find K, therefore it is convenient to first outline how the linear regulator problem is used to feedback stabilize the single control process (2.1).

Introduce the functional

$$I(x,u) = \int_0^\infty \big(\langle q(t)x, x\rangle + \langle r(t)u, u\rangle\big)\,dt,$$

where $u(\cdot)$ is a square-integrable control defined on $[0,\infty)$. Let $x(t)$ be the corresponding solution of (2.1) which satisfies $x(0) = x_0$. The vector $x_0 \in \mathbf{R}^n$ is for the time being fixed. The matrix functions $q(\cdot)$, $r(\cdot)$ are of sizes $n \times n$ and $m \times m$ respectively and are symmetric. One assumes that $q(t) \ge 0$ and $r(t) > 0$ for all $t \in \mathbf{R}$. Finally we assume that q and r are bounded and uniformly continuous in t.

Following standard procedure, we introduce a variable y conjugate to x and define the Hamiltonian

$$H(t,x,y,u) = \langle x', y\rangle + \frac{1}{2}\{\langle q(t)x, x\rangle + \langle r(t)u, u\rangle\}.$$

Inserting the relation $x' = ax + bu$, we obtain

$$H(t,x,y,u) = \langle ax + bu, y\rangle + \frac{1}{2}\{\langle q(t)x, x\rangle + \langle r(t)u, u\rangle\}.$$

Note that $D_uH = ru+b^*y$ and $D_u^2H = r$. Since r is positive definite and hence invertible, the Hamiltonian is convex and regular.

At this point we invoke Pontryagin's Principle [PBGM]: a necessary condition on any control u minimizing the functional I is

$$D_uH = 0.$$

This formula yields immediately the **feedback rule**

$$u = -r^{-1}(t)b^*(t)y. \tag{7.2}$$

Substituting $u = -r^{-1}b^*y$ in the formula for H, we write down Hamilton's equations $x' = H_y$, $y' = -H_x$:

$$\begin{pmatrix} x \\ y \end{pmatrix}' = \begin{pmatrix} a & -br^{-1}b^* \\ -q & -a^* \end{pmatrix} \begin{pmatrix} x \\ y \end{pmatrix}. \tag{7.3}$$

One can now use general arguments involving the regularity and convexity of H to show that the unique control u which minimizes I is given by the feedback rule (7.2), where $y(t)$ is obtained from a certain trajectory $\begin{pmatrix} x(t) \\ y(t) \end{pmatrix}$ of (7.3) satisfying $x(0) = x_0$.

The problem now is to determine $y(t)$, and then to show that the feedback formula (7.2) can be used to stabilize (2.1). The well-known way to achieve these aims is to solve a Riccati equation defined by (7.3). Namely, write $y = m(t)x$ where $m(\cdot)$ is an $n \times n$ -matrix valued function. One obtains the following equation for m:

$$m' = -q - (a^*m + ma) + mbr^{-1}b^*m. \tag{7.4}$$

One seeks a bounded, symmetric, positive-definite solution $m(t) = m^*(t) > 0$ of (7.4). The following controllability conditions ensures that this can be done.

$$x' = a(t)x + b(t)u \qquad \text{is uniformly controllable, and} \tag{7.5.a}$$

$$y' = -a^*(t)y + \sqrt{q(t)}u \qquad \text{is uniformly controllable.} \tag{7.5.b}$$

Here $\sqrt{q(t)}$ stands for the unique positive semi-definite square root of matrix $q(t)$. See [Bo1], [IMK], [K].

Assuming that (7.5.a, b) hold, and having obtained the solution $(x(t), y(t))$ of (7.3) such that $x(0) = x_0$, (where $y(t) = m(t)x(t)$), one verifies that the control $u(t)$ from (7.2) and the function $x(t)$ indeed minimize I. Furthermore, one checks that, if

$$k(t) = -r^{-1}(t)b^*(t)m(t),$$

then $x = 0$ is an exponentially asymptotically stable solution of (7.1).

We now proceed to point out how the analysis involving the Riccati equation can be replaced by arguments involving exponential dichotomy and the rotation number. As has been already pointed out, this approach yields very good properties of robustness, smoothness, and conservation of recurrence for the feedback $k(t)$.

First we must randomize the linear regulator problem. Let $(\Omega, \{T_t\}_{t\in\mathbf{R}})$ be a flow on a compact metric space Ω. Suppose that $Q : \Omega \to M(n,n)$ and $R : \Omega \to M(m,m)$ are continuous functions. Let Q and R satisfy the following condition:

$$Q^*(\omega) = Q(\omega) \geq 0 \qquad \text{for all } \omega \in \Omega; \text{ and}$$
$$R^*(\omega) = R(\omega) > 0 \qquad \text{for all } \omega \in \Omega.$$

We seek to minimize the functional

$$I_\omega(x,u) = \int_0^\infty \big(\langle Q(T_t(\omega))x, x\rangle + \langle R(T_t(\omega))u, u\rangle\big)\,dt$$

for each $\omega \in \Omega$, where now $u(\cdot)$ is a square-integrable function defined on $[0,\infty)$ and $x(t)$ is the corresponding solution of $(2.1.\omega)$ with $x(0) = x_0$.

Let us remark that, if $q(\cdot)$ and $r(\cdot)$ both belong to the class $\mathcal{D}$ of Section 2, then a compact metric space Ω and mappings Q and R as above may be found such that, for some $\omega \in \Omega$,

$$q(t) = Q(T_t(\omega)) \quad \text{and} \quad r(t) = R(T_t(\omega)).$$

Having randomized the quantities q, r and I, we obtain a corresponding randomization of the Hamiltonian H. Following the reasoning outlined above, we are led to the random version of the feedback law (7.2):

$$u = -R^{-1}(T_t(\omega))B^*(T_t(\omega))y, \tag{7.6}$$

and hence to the following random family of Hamiltonian linear equations:

$$\begin{pmatrix} x \\ y \end{pmatrix}' = \begin{pmatrix} A & -BR^{-1}B^* \\ -Q & -A^* \end{pmatrix} \begin{pmatrix} x \\ y \end{pmatrix} \tag{7.7}$$

where all arguments are evaluated at $T_t(\omega)$.

Our program now is as follows. We will show that equations (7.7) have an exponential dichotomy if the following analogues of the controllability conditions $(7.5.a, b)$ are verified:

(7.8.a) $\qquad x' = A(T_t(\omega))x + B(T_t(\omega))u \qquad$ is controllable and

(7.8.b) $\qquad y' = -A^*(T_t(\omega))y + \sqrt{Q(T_t(\omega))}u \qquad$ is controllable,

for each $\omega \in \Omega$. (We will see that this assumption actually implies **uniform** controllability for all $\omega \in \Omega$.) We prove that (7.7) has an exponential dichotomy by using the rotation number; this quantity will be defined and its properties discussed in the next section. Then, we will use the exponential dichotomy property and the controllability conditions $(7.8.a, b)$ to solve the random linear regulator problem and the random feedback stabilization problem.

Before embarking on this program, we notice that controllability for each $\omega \in \Omega$ in $(7.8.a, b)$ actually implies uniform controllability. This follows from Proposition 4.5.

§8. The Rotation Number

Now we define the rotation number of the family of linear Hamiltonian systems given by equations (7.7). We need some auxiliary concepts to do so. To begin, let

$$J = \begin{pmatrix} 0 & -I \\ I & 0 \end{pmatrix}$$

be the standard $2n \times 2n$ skew-symmetric matrix; here I is the $n \times n$ identity matrix. Let $l \subseteq \mathbf{R}^{2n}$ be a vector subspace of dimension n. We say that l is a **Lagrange subspace** (or a **Lagrange plane**) if for each pair of vectors $u, v \in l$, there holds

$$\langle u, Jv \rangle = 0.$$

In more formal language, l is a maximal isotropic subspace of $\mathbf{R}^{2n}$ with respect to the skew 2-form $(u, v) \to \langle u, Jv \rangle$.

Next let $\mathcal{L}$ be the set of all Lagrange planes in $\mathbf{R}^{2n}$. It can be shown that $\mathcal{L}$ is a compact C^∞ manifold of dimension $\frac{n(n+1)}{2}$. In fact, an open dense subset of $\mathcal{L}$ can be parametrized in the following way. Let m be a real, $n \times n$ symmetric matrix with columns $m_1, \cdots, m_n$. Let $e_1, \cdots, e_n$ be the standard basis of $\mathbf{R}^n$. Form n vectors $v_1, \cdots, v_n$ in $\mathbf{R}^{2n}$ by

$$v_1 = \begin{pmatrix} e_1 \\ m_1 \end{pmatrix}, \cdots, v_n = \begin{pmatrix} e_n \\ m_n \end{pmatrix}.$$

Let $l \equiv l_m = \mathrm{Span}\{v_1, \cdots, v_n\}$. It can be checked that l is a Lagrange plane. Further, as m varies over the set of real symmetric $n \times n$ matrices, the Lagrange plane $l \equiv l_m$ traces out an open dense subset $\mathcal{L}_1$ of $\mathcal{L}$. Now we define the **Maslov cycle** $\mathcal{C}$ in $\mathcal{L}$. Let $l_0 = \mathrm{Span}\{\begin{pmatrix} 0 \\ e_1 \end{pmatrix}, \cdots, \begin{pmatrix} 0 \\ e_n \end{pmatrix}\} \subset \mathbf{R}^{2n}$. Then l_0 is a Lagrange plane. Define

$$\mathcal{C} = \{ l \in \mathcal{L} \mid l \cap l_0 \neq \{0\} \}.$$

Thus $\mathcal{C}$ is the set of Lagrange planes which intersect the fixed Lagrange plane l_0 non-trivially. It can be seen, also, that $\mathcal{C}$ is nothing other than the complement $\mathcal{L} - \mathcal{L}_1$ of $\mathcal{L}_1$ in $\mathcal{L}$.

It can be shown (see [Ar]) that $\mathcal{C}$ is a $\mathbf{Z}_2$-cycle of codimension 1 in $\mathcal{L}$. It is not orientable if n is even, but it is **two – sided** in the sense that there is a smooth non-vanishing vector field defined on a neighbourhood of $\mathcal{C}$ in $\mathcal{L}$. This allows one to define an oriented intersection number between $\mathcal{C}$ and any closed curve $c : [0, 1] \to \mathcal{L}$. At a given point of intersection $p = c(t_0) \in \mathcal{C}$, this oriented intersection number takes value between $-n$ and n. See [Ar] for details.

The situation is quite simple if $n = 1$. In this case a Lagrange plane is just a line through the origin in $\mathbf{R}^2$. The set $\mathcal{L}$ is the space of all such lines through the origin, i.e. $\mathcal{L} = \mathbf{P}^1(\mathbf{R})$, the real projective one-dimensional space. Thus $\mathcal{L}$ is homeomorphic to the circle. The Maslov cycle consists of the point p in $\mathcal{L}$ determined by the line containing

$\begin{pmatrix} 0 \\ 1 \end{pmatrix} \in \mathbf{R}^2$. Clearly a closed curve $c : [0,1] \to \mathcal{L}$ which intersects $\mathcal{C}$ in one point will have intersection number $-1, 0$ or 1.

Return now to the set of differential equations (7.7). For fixed $\omega \in \Omega$, let $\Phi_\omega(t)$ be the fundamental matrix solution satisfying $\Phi_\omega(0) = I$. Then $\Phi_\omega(t)$ is a **symplectic** matrix. This means that $\Phi_\omega(t)$ preserves the skew- form defined by J. In fact it can be checked that

$$\langle \Phi_\omega(t)u, J\Phi_\omega(t)v \rangle = \langle u, Jv \rangle \tag{8.1}$$

for all vectors $u, v \in \mathbf{R}^{2n}$.

Now, the linearity of $\Phi_\omega(t)$ together with (8.1) imply that, for fixed ω and t, $\Phi_\omega(t)$ defines a mapping of $\mathcal{L}$ to itself. In fact, if l is a Lagrange plane and $u, v \in l$, then (8.1) implies that

$$\Phi_\omega(t) \cdot l = \{\Phi_\omega(t)v \mid v \in l\}$$

is again a Lagrange plane in $\mathbf{R}^{2n}$.

Let us fix a Lagrange plane $l_1 \notin \mathcal{C}$. Consider the curve $c : [0,T] \to \mathcal{L}$ given by $c(t) = \Phi_\omega(t) \cdot l_1$ where T is a positive number. Assuming that $c(T) \notin \mathcal{C}$, we can deform c through $\mathcal{L}_1 = \mathcal{L} - \mathcal{C}$ so as to get a closed curve $\tilde{c}$ with the same number of points of intersection as c. This is because $\mathcal{L}_1$, being parametrized by the set of $n \times n$ symmetric real matrices, is homeomorphic to a Euclidean space.

Let us now define $n(T)$ to be the sum of the oriented intersection numbers of all points of intersection of $\tilde{c}$ with the Maslov cycle $\mathcal{C}$. We "define" a rotation number for (7.7) as follows:

$$\alpha = \lim_{T \to \infty} \frac{n(T)}{T}. \tag{8.2}$$

That is, α is the average number of times the curve $c = c_T : [0,T] \to \mathcal{L}$ intersects the Maslov cycle, as $T \to \infty$.

There are several obvious difficulties with (8.2). First of all, $c(T)$ may lie on $\mathcal{C}$. However we can perturb c slightly and obtain a well-defined number $n(T)$; it is clear that the limit in (8.2) will not depend on the perturbation. More fundamental is the question of the **existence** of the limit in (8.2). Indeed it need not exist for all $\omega \in \Omega$. However one has the following result [J4].

8.1 Theorem. *Let μ be an ergodic measure on Ω. Then there exists a set $\Omega_1 \subseteq \Omega$ such that $\mu(\Omega - \Omega_1) = 0$ with the following property. If $\omega \in \Omega_1$ and $l \in \mathcal{L}$, then the limit in (8.2) exists and is independent of the choice of $(\omega, l) \in \Omega_1 \times \mathcal{L}$.*

This constant limit is called the **rotation number** of equations (7.7) with respect to the ergodic measure μ.

Theorem 8.1 is true in the following general context. Let $\mathcal{G} : \Omega \to M(2n, 2n)$ be a matrix-valued function of the form

$$\mathcal{G}(\omega) = \begin{pmatrix} g_{11} & g_{12} \\ g_{21} & -g_{11}^* \end{pmatrix}$$

where g_{12} and g_{21} are symmetric $n \times n$ matrices. Thus for each $\omega \in \Omega$ the matrix $\mathcal{G}(\omega)$ is **infinitesimally symplectic**: Let $\mathcal{G}$ satisfy the same conditions as the matrix function A introduced earlier in Section 2. Consider the (Hamiltonian) linear differential equations

$$\begin{pmatrix} x \\ y \end{pmatrix}' = \mathcal{G}(T_t(\omega)) \begin{pmatrix} x \\ y \end{pmatrix}. \tag{8.3}$$

Then the rotation number α for equations (8.3) may be defined by (8.2), and the limit exists and is constant in the sense of Theorem 8.1.

We will see shortly that Theorem 8.1 is rather trivial for equations (7.7), because of the structure of (7.7) and the controllability conditions (7.8.a, b). However, for the general system (8.3), more work is necessary to prove Theorem 8.1.

We can now relate the concept of rotation number and exponential dichotomy. It is convenient to do so for the general Hamiltonian linear equations (8.3). Introduce a continuous, symmetric, positive semi- definite matrix function

$$\Gamma : \Omega \to M(2n, 2n);$$

thus $\Gamma^*(\omega) = \Gamma(\omega) \geq 0$. Suppose that the following "Atkinson condition" [At] is satisfied:

$$\text{for every } \omega \in \Omega, \quad \int_0^\infty ||\Gamma(T_t(\omega))\Phi_\omega(t)v||^2 dt > 0 \text{ for each } v \in \mathbf{R}^{2n} - \{0\}. \tag{8.4}$$

Here $\Phi_\omega(t)$ is the fundamental matrix solution of (8.3) which satisfies $\Phi_\omega(0) = I$.

8.2 Theorem. *Let λ be a real parameter, and consider the random Hamiltonian equations*

$$\begin{pmatrix} x \\ y \end{pmatrix}' = \left[\mathcal{G}(T_t(\omega)) + \lambda J \Gamma(T_t(\omega))\right] \begin{pmatrix} x \\ y \end{pmatrix}. \tag{8.5}$$

Let μ be an ergodic measure on Ω whose topological support is all of Ω. Suppose that the rotation number $\alpha = \alpha(\lambda)$ of (8.5) *(taken with respect to the measure μ) is constant for all λ in some open interval (λ_1, λ_2). Suppose further that the Atkinson condition* (8.4) *is satisfied by $\mathcal{G}$ and Γ. Then for each $\lambda \in (\lambda_1, \lambda_2)$, equations* (8.5) *admit an exponential dichotomy.*

This theorem is proven in [JN3]. The details involve the spectral theory of equation (8.5) discussed in Chapter 9 of Atkinson's book [At] and it would take us too far a field to review them here.

§9. The Solution of The Linear Regulator and The Stabilization Problem.

We wish now to apply Theorem 8.2 to the random family (7.7). Write

$$\mathcal{G} = \begin{pmatrix} A & -BR^{-1}B^* \\ -Q & -A^* \end{pmatrix},$$

and introduce a positive semi-definite, matrix valued function Γ by

$$\Gamma(\omega) = \begin{pmatrix} Q(\omega) & 0 \\ 0 & BR^{-1}B^*(\omega) \end{pmatrix} \qquad (\omega \in \Omega).$$

Let μ be an ergodic measure on Ω. We are going to show that Theorem 8.2 applies to the random family

$$\begin{pmatrix} x \\ y \end{pmatrix}' = [\begin{pmatrix} A & -BR^{-1}B^* \\ -Q & -A^* \end{pmatrix} + \lambda J \begin{pmatrix} Q & 0 \\ 0 & BR^{-1}B^* \end{pmatrix}] \begin{pmatrix} x \\ y \end{pmatrix}, \tag{9.1}$$

where $-\frac{1}{2} < \lambda < \frac{1}{2}$. We first verify that the assumptions of constancy of the rotation number α is satisfied. Indeed we will show that $\alpha(\lambda) = 0$ if $-\frac{1}{2} < \lambda < \frac{1}{2}$.

9.1 Proposition. *The rotation number* $\alpha = \alpha(\lambda, \mu) = 0$ *for equations* (9.1), *for each* $\lambda \in (-\frac{1}{2}, \frac{1}{2})$ *and for each ergodic measure* μ *on* Ω.

Proof. Let $T > 0$. We consider the boundary value problem consisting of equations (9.1) together with the boundary conditions

$$x(0) = x(T) = 0.$$

We will show first that this boundary value problem has only the trivial solution if T is sufficiently large and $-\frac{1}{2} < \lambda < \frac{1}{2}$.

To do this, let $\begin{pmatrix} x(t) \\ y(t) \end{pmatrix}$ be a solution of (9.1) such that $x(0) = x(T) = 0$. Then

$$\begin{aligned} 0 &= \langle x(T), y(T)\rangle - \langle x(0), y(0)\rangle \\ &= \int_0^T \frac{d}{dt}\langle x(t), y(t)\rangle dt \\ &= \int_0^T [\langle \frac{dx}{dt}, y\rangle + \langle x, \frac{dy}{dt}\rangle] dt \\ &= \int_0^T [\langle Ax - (\lambda+1)BR^{-1}B^*y, y\rangle + \langle x, (\lambda-1)Qx - A^*y\rangle] dt. \end{aligned}$$

It follows that

$$0 = (\lambda - 1)\int_0^T ||Cx||^2 dt - (\lambda+1)\int_0^T ||R^{-1/2}B^*y||^2 dt,$$

where $C(\omega) = \sqrt{Q(\omega)}$ (the unique positive semi-definite square root). Thus

$$(\lambda - 1)\int_0^T ||Cx||^2 dt = (\lambda+1)\int_0^T ||R^{-1/2}B^*y||^2 dt. \tag{9.2}$$

Now, since $-\frac{1}{2} < \lambda < \frac{1}{2}$, we must have $C(T_t(\omega))x(t) = 0 = B^*(T_t(\omega))$ for all $0 \le t \le T$. We conclude that $y'(t) = -A^*(T_t(\omega))y(t)$. Using the controllability of $y' = Ay + Bu$, (7.8.b) and the relation

$$\int_0^T ||B^* y||^2 dt = 0,$$

we see that $y(t) = 0$ on $[0, T]$ if T is sufficiently large, i.e., if T is large enough so that the controllability matrix of $y' = Ay + Bu$ is strictly positive definite. Since $x(0) = 0$, we have that $x(t)$ is identically zero on $[0, T]$, as well.

Next let $l_0 = \{\begin{pmatrix} 0 \\ y \end{pmatrix} \mid y \in \mathbf{R}^n\}$ be the Lagrange plane used in defining the Maslov cycle $\mathcal{C}$. Write $\Phi_\lambda(\omega, t)$ for the fundamental matrix solution of (9.1) which satisfies $\Phi_\lambda(\omega, t) = I$. Note that, if

$$\dim\ (\Phi_\lambda(\omega, T)l_0 \cap l_0) \ge 1$$

for some $T > 0$ and some $\lambda \in (-\frac{1}{2}, \frac{1}{2})$, then there exists $y_0 \in \mathbf{R}^n - \{0\}$ such that

$$\Phi_\lambda(\omega, T) \begin{pmatrix} 0 \\ y_0 \end{pmatrix} = \begin{pmatrix} 0 \\ y(T) \end{pmatrix}.$$

But then by the preceding considerations $y_0 = 0$ if T is large enough, a contradiction.

This shows that, if T is large enough and $\lambda \in (-\frac{1}{2}, \frac{1}{2})$, then $\Phi_\lambda(\omega, T)l_0$ does not lie on the Maslov cycle. A brief review of the definition of the rotation number shows that $\alpha(\lambda)$ must be zero, for division by T in the limit (8.2) causes that limit to be zero. This proves Proposition 9.1. $\square$

To apply Theorem 8.2 to the random family (9.1), we must still verify the Atkinson condition (8.4). This condition is in fact equivalent to the controllability condition $(7.8.a, b)$. We omit the details of the proof; they are given in [JN5].

At this point we can use Theorem 8.2 to conclude that equations (7.7) have an exponential dichotomy over the topological support of the measure μ (just set $\lambda = 0$ in equations (9.1)). We wish, however, to conclude that equations (7.7) have an exponential dichotomy over all of Ω. For this, we need some additional reasoning which we now outline.

Let M be a minimal subset of Ω. Then M is always the topological support of an ergodic measure μ. This follows from minimality of M, the classical Krylov-Bogoliubov construction on invariant measure on M [NS], and the Krein-Milman theorem, (see, e.g., [JPS]). Hence Theorem 8.2 implies that system (7.7) has an exponential dichotomy over M.

Next let $P = P(\omega) : \mathbf{R}^{2n} \to \mathbf{R}^{2n}$ be the projection in the definition of exponential dichotomy (see (3.2)). Since (7.7) is Hamiltonian, one must have $\dim(\text{Im } P(\omega)) = n$; indeed Im $P(\omega) \subset \mathbf{R}^{2n}$ is a Lagrange plane [J4]. Thus equations (7.7) have exponential dichotomy over each minimal subset $M \subseteq \Omega$, and the dimension of the image of $P(\omega)$ is constant independent of M.

Now, a criterion of Sacker and Sell [SS] states that, if the above conditions hold, and if (7.7) admits no solution bounded on $-\infty < t < \infty$ for any $\omega \in \Omega$, then equations (7.7) indeed have an exponential dichotomy. To see that none of the equations (7.7) admit a bounded solution fix $\omega \in \Omega$, let $-\infty < S < T < \infty$, and consider the equations

$$\begin{aligned}\langle x(T), y(T)\rangle - \langle x(S), y(S)\rangle &= \int_S^T \frac{d}{dt}\langle x(t), y(t)\rangle dt \\ &= \int_S^T \left[\langle Ax - BR^{-1}B^*y, y\rangle - \langle x, Qx + A^*y\rangle\right] dt,\end{aligned}$$

which are valid for any solution $\begin{pmatrix} x(t) \\ y(t) \end{pmatrix}$ of (7.7). Hence

$$\langle x(T), y(T)\rangle - \langle x(S), y(S)\rangle = -\int_S^T ||Cx||^2 dt - \int_S^T ||R^{-1/2}B^*y||^2 dt, \tag{9.3}$$

where $C = \sqrt{Q}$.

Next note that there exist sequences $T_n \to \infty$, $S_n \to -\infty$ such that $\begin{pmatrix} x(T_n) \\ y(T_n) \end{pmatrix} \to \begin{pmatrix} 0 \\ 0 \end{pmatrix}$ and $\begin{pmatrix} x(S_n) \\ y(S_n) \end{pmatrix} \to \begin{pmatrix} 0 \\ 0 \end{pmatrix}$. For if, for example, there is no such sequence S_n, then one verifies that, for each $\overline{\omega}$ in the α-limit set of ω, the equation (7.7) corresponding to $\overline{\omega}$ admits a non-trivial solution bounded on all of $\mathbf{R}$. But the α-limit set of ω contains a minimal subset $M \subseteq \Omega$. Since equations (7.7) have an exponential dichotomy over M, they do not admit bounded solutions there. So the sequences $\{T_n\}, \{S_n\}$ must exist.

But now (9.3) implies

$$0 = \lim_{n\to\infty} \langle x(T_n), y(T_n)\rangle - \langle x(S_n), y(S_n)\rangle = -\int_{-\infty}^{\infty} \left[\, ||Cx||^2 + ||R^{-1/2}B^*y||^2\right] dt.$$

Using the controllability conditions (7.8.a, b) and arguing as in the proof of Proposition 9.1, we see that $x(t) = y(t) = 0$ for all $t \in \mathbf{R}$.

We conclude that the Sacker-Sell criterion is applicable, and hence equations (7.7) admit an exponential dichotomy over all of Ω.

We finish this section by showing how to solve the linear regulator and feedback control problems. Consider the projection $P = P(\omega)$ for fixed $\omega \in \Omega$. If a vector $\begin{pmatrix} x_0 \\ y_0 \end{pmatrix}$ lies in Im P, then Definition 3.2 states that the solution $\begin{pmatrix} x(t) \\ y(t) \end{pmatrix}$ of (7.7) with $\begin{pmatrix} x(0) \\ y(0) \end{pmatrix} = \begin{pmatrix} x_0 \\ y_0 \end{pmatrix}$ tends to zero exponentially as $t \to \infty$. Referring to (7.6), we see that then $u(t) = -R^{-1}(T_t(\omega))B^*(T_t(\omega))y(t)$ is certainly a control for which the functional $I(x, u)$ is finite.

We want to show that (7.6) yields a control for which $I(x,u) < \infty$ for all initial vectors $x_0 \in \mathbf{R}^n$. This amounts to requiring that Im $P(\omega)$ be a graph over $\mathbf{R}^n$. Precisely, we require the existence of an $n \times n$ matrix $m = m(\omega)$ such that

$$\text{Im } P(\omega) = \{\begin{pmatrix} x \\ m(\omega)x \end{pmatrix} \mid x \in \mathbf{R}^n\}.$$

Let us show that such a matrix m exists. But first note that, if m does exist, then it is necessarily symmetric (because Im P is a Lagrange plane, as noted earlier). Now, the existence of m is equivalent to the non-existence of a vector of the form $\begin{pmatrix} 0 \\ y \end{pmatrix}$ in Im P with $y \neq 0$. To show that there is no such vector, we appeal once more to the controllability conditions (7.8.a, b). We in fact show that, if $\begin{pmatrix} x_0 \\ y_0 \end{pmatrix}$ is in Im P, then the inner product $\langle x_0, y_0 \rangle \neq 0$ unlesss $x_0 = y_0 = 0$. To see this, suppose $\langle x_0, y_0 \rangle = 0$, and let $\begin{pmatrix} x(t) \\ y(t) \end{pmatrix}$ be the corresponding solution of (7.7). Then

$$\begin{aligned}\langle x(T), y(T) \rangle &= \int_0^T \frac{d}{dt} \langle x(t), y(t) \rangle dt \\ &= -\int_0^T [\ ||Cx||^2 + ||R^{-1/2}B^*y||^2] dt.\end{aligned}$$

Since $\begin{pmatrix} x_0 \\ y_0 \end{pmatrix} \in$ Im P, we have $x(T) \to 0,\ y(T) \to 0$ as $T \to \infty$. Hence

$$0 = C(T_t(\omega))x(t) = R^{-1/2}(T_t(\omega))B^*(T_t(\omega))y(t) \quad \text{for all } t \in \mathbf{R}.$$

Using the controllability conditions (7.8.a, b) and arguing once more as in the proof of Proposition 9.1, we conclude that $x(t) = 0 = y(t)$ for all $t \in \mathbf{R}$. In particular $x_0 = y_0 = 0$.

We see that, for all $\omega \in \Omega$, we have

$$\text{Im } P(\omega) = \{\begin{pmatrix} x \\ m(\omega)x \end{pmatrix} \mid x \in \mathbf{R}^n\},$$

where $m(\omega) = m^*(\omega)$. The continuity of $P(\cdot)$ implies the continuity in ω of m. This is the source of the conservation of recurrence discussed earlier. One verifies that, for each $\omega \in \Omega$, the function $t \to m(T_t(\omega))$ is a symmetric solution of the Riccati equation (7.4). Since $\langle x, m(\omega)x \rangle \neq 0$ for all $x \in \mathbf{R}^n$, $m(\omega)$ is also a definite matrix; we shall see shortly that it is **positive definite**.

We can now solve the random linear regulator problem. For fixed $\omega \in \Omega$, let $\tilde{m}(t)$ be any symmetric solution of the Riccati equation (7.4). Let $u(t)$ be an $\mathbf{R}^m$- valued locally integrable function defined on $[0, \infty)$ and let $x(t)$ be the solution of

$$x' = A(T_t(\omega))x + B(T_t(\omega))u,$$

satisfying $x(0) = x_0$. By direct calculation one verifies the following identity:

$$\frac{d}{dt}\langle \tilde{m}(t)x(t), x(t)\rangle = ||R^{1/2}(T_t(\omega))\big[u(t) + R^{-1}B^*(T_t(\omega))\tilde{m}(t)x(t)\big]||^2 - \big[\langle Q(T_t(\omega))x, x\rangle + \langle R(T_t(\omega))u, u\rangle\big].$$

Assuming that $I(x,u)$ is finite and that $x(t) \to 0$ as $t \to \infty$, we must have

$$I(x,u) = \langle \tilde{m}(0)x_0, x_0\rangle + \int_0^\infty ||R^{1/2}\big[u + R^{-1}B^*\tilde{m}x\big]||^2 dt. \tag{9.4}$$

Now (in accordance with the feedback rule (7.6)) choose

$$u(t) = -R^{-1}(T_t(\omega))B^*(T_t(\omega))y(t) \text{ and}$$
$$y(t) = m(T_t(\omega))x(t)$$

where $\begin{pmatrix} x(t) \\ y(t) \end{pmatrix}$ solves (7.7). In this case (9.4) becomes

$$I_\omega(x,u) = \langle m(\omega)x_0, x_0\rangle.$$

Since I_ω is non-negative, we conclude that $m(\omega)$ is positive semi-definite, and hence positive definite using the second paragraph above. Now we appeal to the standard arguments which show that $u(t)$ and $x(t)$ define the unique minimizer (x,u) of I_ω.

We can now solve the random feedback stabilization problem, as well. We have merely to choose continuous functions Q, R which satisfy the controllability conditions (7.8.a,b). We can, for instance, choose Q and R to be constant symmetric positive definite matrices. In any case, writing

$$K(\omega) = -R^{-1}(\omega)B^*(\omega)m(\omega), \tag{9.5}$$

and substituting $u(t) = K(T_t(\omega))x(t)$ in to equation (2.1.ω), we see that (2.1.ω) is indeed exponentially stabilized.

Let us briefly discuss the smoothness properties of K. First of all, suppose that A and B are C^r-smooth in parameter $\lambda \in \mathbf{R}^p$, and that their derivatives of order up to and including r are jointly continuous in (λ, ω). Choose Q and R to be constant, positive definite matrices. Then equations (7.7) have an exponential dichotomy, and results of Palmer [P3] and Yi [Y] may be used to show that the function $m = m(\omega, \lambda)$ enjoys these properties as well.

Next suppose Ω is a C^r-manifold. In this case, $m(\cdot)$ need not be C^r in ω. However, let N be the "largest Lyapunov exponent" of the flow $(\Omega, \{T_t\})$. More precisely, let d be a metric on Ω, and let N be a number such that

$$d(T_t(\omega_1), T_t(\omega_2)) \le d(\omega_1, \omega_2)e^{N|t|} \text{ for all } t \in \mathbf{R},\ \omega_1, \omega_2 \in \Omega,$$

and let α be the constant appearing in the definition of exponential dichotomy (Definition 3.2). If

$$(r+1)N < \frac{\alpha}{2},$$

then a result of [Y] implies that m is C^r-smooth in ω. Now, the fact is that Q and R can be chosen so as to make α as large as desired [JN5]. So with an appropriate choice of Q and R, we can obtain a feedback matrix $K(\omega)$ which is C^r-smooth in ω.

It should be also noted that the exponential rate of stabilization can be made as large as desired. That is, given a positive constant c, we can determine (constant) matrices Q and R in such a manner that, for some constant $c_1 > 0$,

$$||x(t)|| \leq c_1 e^{-ct} ||x(0)||$$

for all solutions $x(t)$ of $x' = [A+BK]x$. This is the random analogue of "pole relocation", and its proof again involves arranging that the dichotomy constant α be sufficiently large. See again [JN5].

Let us summarize sections 7,8 and 9. We have produced a feedback matrix $K(\omega)$ which is continuous in ω, so that conservation of recurrence holds for the random feedback stabilization problem. Moreover K is smooth in parameters if A and B are, and is smooth in ω if Ω is a smooth manifold and $A(\cdot)$, $B(\cdot)$ are smooth. In addition K is directly related to the dichotomy projection P, and so all the results in the literature concerning robustness of the dichotomy projections carry over to K as well (e.g. [Cp], [SS]).

§10. Linearization of The Regulator and The Stabilization Problem.

In this section we show how the linear results of the preceding sections can be used to solve the non-linear feedback control problem in a random context. We will then formulate and solve the random non-linear regulator problem, as well. As we shall see, the two problems can be treated independently once the linear results of Section 9 are available.

We begin by considering a typical situation in which the random feedback control problem arises. Consider the control problem

$$(10.1) \qquad x' = f(x,u) \qquad x \in \mathbf{R}^n, \; u \in \mathbf{R}^m$$

where $f : \mathbf{R}^n \times \mathbf{R}^m \to \mathbf{R}^n$ is a function of class C^2. Suppose that the equation

$$(10.2) \qquad x' = f(x,0)$$

admits a compact invariant set $\Omega \subset \mathbf{R}^n$. This set could be a fixed point, in which case one has the well-studied problem of controlling a non-linear system near an equilibrium point. See for example the book of A. Bacciotti [B] for recent results on this topic. Or, Ω could be a periodic orbit, in which case the feedback stabilization problem can be studied using the well-developed theory of periodic differential systems.

Our formulation includes **all** possibilities when Ω is compact. For example, Ω could be a compact manifold, and the vector field $x \to f(x,0)$ might generate a chaotic flow. As another example, Ω might be a topological k-torus and the flow on Ω defined by f might be quasi-periodic, that is up to a homeomorphism it is of the form

$$T_t(\omega_1, \cdots, \omega_k) = (\omega_1 + \gamma_1 t, \cdots, \omega_k + \gamma_k t)$$

where $\omega = (\omega_1, \cdots, \omega_k) \in \Omega$ and $(\gamma_1, \cdots, \gamma_k)$ is a frequency vector which does not depend on time.

We pose the problem, then, of studying the control problem (10.1) in a neighbourhood of Ω. To do so, we first linearize (10.1) in a neighbourhood of Ω. This is done as follows. For fixed $\omega \in \Omega$ we write

$$x = \omega + y$$

where $x \in \mathbf{R}^n$, and compute

$$y' = A(T_t(\omega))y + B(T_t(\omega))u + C((T_t(\omega))yu + F(T_t(\omega), y, u), \tag{10.3}$$

where $F(\omega, y, u) = o(|y|^2 + |u|^2)$ as $(y, u) \to (0, 0)$. Here

$$A(\omega) = \frac{\partial f}{\partial x}(\omega, 0), \quad B(\omega) = \frac{\partial f}{\partial u}(\omega, 0), \quad C(\omega) = \frac{\partial^2 f}{\partial y \partial u}(\omega, 0).$$

It will sometimes be convenient to write $\mathcal{G}(T_t(\omega), y, u)$ for the right hand side in (10.3); thus

$$\mathcal{G}(\omega, y, u) = A(\omega)y + B(\omega)u + C(\omega)yu + F(\omega, y, u).$$

Our goal is to find a feedback control $u = u(\omega, x)$ which takes solutions $x(t)$ of (10.1) beginning in some neighbourhood V of Ω to the set Ω at an exponential rate. It is easier to make this condition precise making use of the variable y. Thus we require a feedback control $u = u(\omega, y)$ such that $y = 0$ is an exponentially asymptotically stable solution of the system

$$y' = \mathcal{G}\big(T_t(\omega), y, u(T_t(\omega), y)\big), \tag{10.4}$$

uniformly in $\omega \in \Omega$. We further require that u be continuous in ω, and smooth in ω if Ω is a smooth manifold and $\mathcal{G}$ is smooth. If $\mathcal{G}$ depends smoothly on parameters, then u should depend smoothly on those parameters, as well.

This problem is easy to solve with the results of Section 9 and standard stability theory of non-autonomous differential systems. Consider the random equations obtained by linearizing (10.3) around $y = 0$ (which are the same as those obtained by linearizing (10.1) around Ω):

$$y' = A(T_t(\omega))y + B(T_t(\omega))u. \tag{10.5}$$

Assume that the controllability conditions (7.8.a, b) holds. The arguments of Section 4 then show that

$$x' = A(T_t(\omega))x + B(T_t(\omega))u$$

is uniformly controllable for each $\omega \in \Omega$.

As discussed in Section 9, we can find a linear feedback control

$$u(\omega, y) = K(\omega)y \tag{10.6}$$

such that, for each $\omega \in \Omega$,

$$y' = \big[A(T_t(\omega)) + B(T_t(\omega))K(T_t(\omega))\big]y \tag{10.7.ω}$$

admits $y = 0$ as an exponentially, asymptotically stable solution. Precisely: there are constants $L > 0$, $\beta > 0$ such that, if $Y(\omega, t)$ is the fundamental matrix solution of (10.7.ω) satisfying $Y(\omega, 0) = I$-the $n \times n$ identity, then

$$||Y(\omega, t)|| \leq Le^{-\beta(t-s)} \qquad \text{for all } s \leq t \text{ and } \omega \in \Omega.$$

The feedback matrix $K(\cdot)$ is continuous in $\omega \in \Omega$, is smooth in parameters if A and B are smooth in these parameters, and is smooth in ω if Ω is a smooth manifold and the flow on Ω is smooth. Moreover, small perturbations of A and B produce a Lipschitz perturbation of K.

We now plug (10.6) into (10.3), to obtain

$$y' = \big[A(T_t(\omega)) + B(T_t(\omega))K(T_t(\omega))\big]y + o(|y|^2), \tag{10.8}$$

where the non-linear terms are $o(|y|^2)$ uniformly in $\omega \in \Omega$. We now appeal to standard results in dynamical systems theory (see, e.g., [CY]). There is a neighbourhood V of $y = 0$ in $\mathbf{R}^n$ such that, if $y_0 \in V$, then the solution $y(t)$ of (10.8) satisfying $y(0) = y_0$ also satisfies

$$||y(t)|| \leq 2L||y(0)||e^{-\beta/2(t-s)} \qquad \text{for all } s \leq t, \; s, t \in \mathbf{R}.$$

Thus the control $u(t) = K(T_t(\omega))y(t)$ indeed stabilizes equations (10.3), uniformly in $\omega \in \Omega$.

Returning to equations (10.1), we see that the feedback control

$$u(\omega, x) = K(\omega)(x - \omega)$$

stabilizes (10.1) in the following sense. There is a neighbourhood U of Ω in $\mathbf{R}^n$ (namely $U = \{\omega + y \mid \omega \in \Omega, \; y \in V\}$ such that, if $x_0 \in U$, then the solution $x(t)$ of

$$x' = f(x, u(T_t(\omega), x)) \tag{10.9}$$

tends exponentially to Ω as $t \to \infty$. Furthermore, standard theory ([CY]) shows that the following version of the "tracking' property" holds. If $x_0 \in U$ and if $\omega_0 \in \Omega$ is a point sufficiently near to x_0 (in the precise sense that $x_0 - \omega_0 \in V$), then the solution $x(t)$ of (10.7) with $x(0) = x_0$ satisfies

$$||x(t) - T_t(\omega_0)|| \to 0, \quad \text{as } t \to \infty,$$

where the convergence is exponentially fast. We summarize our results in the following theorem.

10.1 Theorem. *Consider the control system (10.1), and let $\Omega \subset \mathbf{R}^n$ be a compact set which is invariant for the system (10.2). Then there exists an $\varepsilon > 0$ and a continuous feedback matrix function $K : \Omega \to M(m,n)$ such that, if $x_0 \in \mathbf{R}^n$ and if $||x_0 - \omega_0|| < \varepsilon$ for some $\omega_0 \in \Omega$, then the linear feedback control*

$$u = K\big(T_t(\omega_0)\big)\big(x - T_t(\omega_0)\big)$$

exponentially stabilizes x_0 to the trajectory through ω_0, i.e., if $x(t)$ is the solution of (10.9) with $x(0) = x_0$ and $\omega = \omega_0$, then $||x(t) - T_t(\omega_0)|| \to 0$ exponentially as $t \to \infty$. If A and B are of class C^r with respect to a parameter $\lambda \in \mathbf{R}^p$, then K is of class C^r in λ. If Ω is a C^r-manifold, then K can be chosen to be C^r in ω. Finally, K is Lipschitz continuous with respect to C^0 perturbations of A and B.

This completes our brief discussion of non-linear feedback stabilization. We next turn to an equally brief discussion of the non-linear regulator problem. We formulate this problem in the following way. Let Ω be a compact metric space and $\{T_t\}_{t\in\mathbf{R}}$ be a flow on Ω. Define the Lagrangian

$$L(x,u,\omega) = \frac{1}{2}\big[\langle x, Q(\omega)x\rangle + \langle u, R(\omega)u\rangle\big] + \hat{L}(x,u,\omega),$$

where $Q(\cdot)$ and $R(\cdot)$ are continuous, symmetric, matrix-valued functions on Ω such that $Q(\omega) \geq 0$ and $R(\omega) > 0$ for each $\omega \in \Omega$. We also require that $\hat{L} : \mathbf{R}^n \times \mathbf{R}^m \times \Omega \to \mathbf{R}$ be jointly continuous in its three arguments and that all of its derivatives of order up to and including three in (x,u) are jointly continuous in (x,u,ω). Furthermore $\hat{L}$ will satisfy

$$D_1\hat{L}(0,0,\omega) = D_1^2\hat{L}(0,0,\omega) = 0, \tag{10.10}$$

$$D_2\hat{L}(0,0,\omega) = D_2^2\hat{L}(0,0,\omega) = D_1D_2\hat{L}(0,0,\omega) = 0, \tag{10.11}$$

for all $\omega \in \Omega$, where here and below D_1 and D_2 denote the gradients with respect to the first component x and the second component u respectively.

Next we introduce a non-linear control process

$$x' = A(T_t(\omega))x + B(T_t(\omega))u + g(x,u,T_t(\omega)), \tag{10.12}$$

where A and B satisfy the requirements placed on the process (2.1.ω) in Section 2. We also require that all derivatives of g in (x,u) of order up to and including 3 are jointly continuous in (x,u,ω). It is further required that

$$g(0,0,\omega) = D_1g(0,0,\omega) = D_2g(0,0,\omega) = 0 \tag{10.13}$$

for all $\omega \in \Omega$. For example, (10.12) might arise from a non-linear process (10.1) via a compact invariant subset Ω of the vector field $x \to f(x,0)$. In that case we would have $g(x,u,\omega) = C(\omega)yu + F(y,u,\omega)$ where $x = \omega + y$.

Now the non-linear quadratic regulator problem can be stated as follows. Given $x_0 \in \mathbf{R}^n$, find a control $u(t) \equiv u(T_t(\omega), x(t))$ with values in $\mathbf{R}^m$ which is square-integrable on $[0, \infty)$ such that, if $x(t)$ is the corresponding solution of (10.12) with $x(0) = x_0$, then the pair $(x(t), u(t))$ minimizes the functional

$$I_\omega(x, u) = \int_0^\infty L(x(t), u(t), T_t(\omega))dt,$$

for all $\omega \in \Omega$.

As it stands this problem need not have a solution, as we have imposed no condition on the non-linear term $\hat{L}(x, u, \omega)$ aside from (10.10) and (10.11). We will content ourselves with finding a control u in feedback form which results in a finite value of $I_\omega(x, u)$ and which satisfies the Pontryagin necessary condition (see below). This control will have good smoothness properties with respect to ω and with respect to parameters, and good robustness properties. One can probably show that the control u defines "locally" a minimum of the functional I_ω, but we prefer not to dwell on this point here.

In any case we follow the standard approach for attacking the non- linear regulator problem. Introduce the Hamiltonian

$$H(x, y, u, \omega) = L(x, u, \omega) + \langle y, A(\omega)x + B(\omega)u + g(x, u, \omega)\rangle. \tag{10.14}$$

We obtain the following formulas for the first and second (Frechét) derivatives of H with respect to u:

$$\frac{\partial H}{\partial u} = Ru + B^* y + (D_2 g)^* y + D_2 \hat{L} \tag{10.14.a}$$

$$\frac{\partial^2 H}{\partial u^2} = R + (D_2^2 g)^* y + D_2^2 \hat{L}. \tag{10.14.b}$$

According to the Pontryagin principle, a necessary condition on an extremizing control $u(t)$ and a corresponding solution $x(t)$ of (10.12) such that $x(0) = x_0$ is

$$\frac{\partial H}{\partial u}(x(t), y(t), u(t), T_t(\omega)) = 0.$$

Here $y(t)$ is obtained from a solution $\begin{pmatrix} x(t) \\ y(t) \end{pmatrix}$ of the Hamiltonian equations (10.16) below. By (10.14.a), the necessary condition is

$$Ru + B^* y + (D_2 g)^* y + D_2 \hat{L} = 0. \tag{10.15}$$

We wish to solve this equation for u in terms of x, y and ω, for x and y near zero. This will give us a feedback formula for u. For this purpose we use the implicit function theorem. Let

$$\Psi(x, y, u, \omega) = Ru + B^* y + (D_2 g)^* y + D_2 \hat{L}.$$

Then $\Psi : \mathbf{R}^{2n} \times \mathbf{R}^m \times \Omega \to \mathbf{R}^m$ is C^2 as a function of (x, y, u), and its derivatives of order up to and including 2 are jointly continuous in (x, y, u, ω). Furthermore

$$\Psi(0,0,0,\omega) = D_2\hat{L}(0,0,\omega) = 0, \text{ and}$$
$$D_3\Psi(0,0,0,\omega) = R, \quad \text{for all } \omega \in \Omega.$$

Now we use positive definiteness of R, continuity of R in ω, compactness of Ω, and the implicit function theorem to conclude that there exists a $\delta > 0$, independent of ω, with the following property. Let $B_\delta(0)$ be the ball centered at the origin in $\mathbf{R}^{2n}$ of radius δ; then there is a C^2 map $F = F(\cdot, \cdot, \omega) : B_\delta(0) \to \mathbf{R}^m$ such that $F(0,0,\omega) = 0$ and

$$\Psi(x, y, F(x, y, \omega), \omega) = 0.$$

We regard the function F as defining a **feedback rule**

$$u(t) = F(x, y, T_t(\omega)).$$

Now substitute this control u in the Hamiltonian H. We obtain a function which by abuse of notation we again call $H : B_\delta(0) \times \Omega \to \mathbf{R}$, and it is given by

$$H(x,y,\omega) = L(x, F(x,y,\omega), \omega) + \langle y, A(\omega)x + B(\omega)F(x,y,\omega)\rangle + \langle y, g(x, F(x,y,\omega), \omega)\rangle.$$

We consider the Hamiltonian equations

$$x' = \frac{\partial H}{\partial y}(x, y, T_t(\omega)), \tag{10.16.a}$$

$$y' = -\frac{\partial H}{\partial x}(x, y, T_t(\omega)), \tag{10.16.b}$$

for each $\omega \in \Omega$. A straight forward verification which we omit, shows that the linear part of the Hamiltonian vector field in (10.16.a, b) is the same as the right-hand side of (7.7):

$$\begin{pmatrix} x \\ y \end{pmatrix} = \begin{pmatrix} A & -BR^{-1}B^* \\ -Q & -A^* \end{pmatrix} \begin{pmatrix} x \\ y \end{pmatrix} \tag{10.17}$$

where all arguments are evaluated at $T_t(\omega)$.

Let us now differentiate the relation $0 = \Psi(x, y, F(x, y, \omega), \omega)$ with respect to x at the point $x = y = 0$. Using the definition of Ψ and condition (10.11), we get

$$0 = D_3\Psi \circ D_x F\big|_{x=y=0},$$

or equivalently $R \circ D_x F(0,0,\omega) = 0$. Since R is positive definite, we have

$$D_x F(0,0,\omega) = 0.$$

In a similar way, differentiation of $\Psi = 0$ with respect to y gives

$$D_y F(0,0,\omega) = R^{-1}(\omega)B^*(\omega).$$

We may therefore write

$$F(x,y,\omega) = R^{-1}(\omega)B^*(\omega)y + \mathcal{G}(x,y,\omega), \tag{10.18}$$

where $\mathcal{G} : B_\delta(0) \times \Omega \to \mathbf{R}^m$ is of class C^2 in (x,y) and all derivatives of order up to and including 2 are jointly continuous in (x,y,ω). We have

$$\mathcal{G}(0,0,\omega) = 0 = D_1\mathcal{G}(0,0,\omega) = D_2\mathcal{G}(0,0,\omega). \tag{10.19}$$

Now put the expression for F in (10.18) into the feedback rule $u(t) = F(x,y,T_t(\omega))$, and put this expression for u into the Hamiltonian H. Equations (10.16.a,b) take the form

$$\begin{pmatrix} x \\ y \end{pmatrix}' = \begin{pmatrix} A & -BR^{-1}B^* \\ -Q & -A^* \end{pmatrix}\begin{pmatrix} x \\ y \end{pmatrix} + \begin{pmatrix} P_1(x,y,\omega) \\ P_2(x,y,\omega) \end{pmatrix}, \tag{10.20}$$

where $P_1(x,y,\omega) = D_y S(x,y,\omega)$, $P_2(x,y,\omega) = D_x S(x,y,\omega)$, and $S : B_\delta(0) \times \Omega \to \mathbf{R}$ is given by

$$\begin{aligned} S(x,y,\omega) = \langle R^{-1}B^*y + \mathcal{G}, R\mathcal{G}\rangle + 2\langle y, B\mathcal{G}\rangle + \hat{L}(x, R^{-1}B^*y + \mathcal{G}, \omega) \\ + \langle y, g(x, R^{-1}B^*y + \mathcal{G}, \omega)\rangle. \end{aligned}$$

Since S is of class C^2, P_1 and P_2 are of class C^1 in x and y, and the derivatives are continuous in (x,y,ω).

At this point we impose the controllability conditions (7.8.a,b). From the previous discussion, we conclude that the linear equations (10.17) have an exponential dichotomy. Let $\alpha > 0$ be the constant appearing in the definition of exponential dichotomy (see Definition 3.2).

We now sketch how, for each $x_0 \in \mathbf{R}^n$ sufficiently near the origin, we can find a unique solution $\begin{pmatrix} x(t) \\ y(t) \end{pmatrix}$ of (10.20) with $x(0) = x_0$ and such that $\begin{pmatrix} x(t) \\ y(t) \end{pmatrix}$ tends to zero exponentially as $t \to \infty$. Moreover $y_0 = y(0)$ is uniquely defined by x_0 and by $\omega \in \Omega$, and depends "nicely" on ω and on parameters as explained below.

First of all, the results of Section 9 imply that the linear subspace

$$l_\omega = \{\begin{pmatrix} x_0 \\ m(\omega)x_0 \end{pmatrix} \mid x_0 \in \mathbf{R}^n\} \subseteq \mathbf{R}^{2n}$$

consists of exactly those points such that the solution $\begin{pmatrix} x(t) \\ y(t) \end{pmatrix}$ of (10.17) with $x(0) = x_0$, $y(0) = y_0 = m(\omega)x_0$ decays exponentially as $t \to \infty$. Since equations (10.17) have an exponential dichotomy, we can use stable manifold theory (see, e.g. [CY] for the results we need) to draw the following conclusions. There is a number $\rho > 0$ and there are C^1-manifolds $V_\omega \subset \{\omega\} \times \mathbf{R}^{2n}$ of dimension n such that, if $\begin{pmatrix} x_0 \\ y_0 \end{pmatrix} \in V_\omega$ then the solution $\begin{pmatrix} x(t) \\ y(t) \end{pmatrix}$ of (10.20) always remains in the ball $B_\rho(0) \subset \mathbf{R}^{2n}$ (of center 0 and radius ρ) for all $t \geq 0$ and tends to zero exponentially as $t \to \infty$. Conversely, if $\begin{pmatrix} x(t) \\ y(t) \end{pmatrix}$ is a solution of (10.20) which remains in $B_\rho(0)$ for all $t \geq 0$, then $\begin{pmatrix} x(0) \\ y(0) \end{pmatrix} \in V_\omega$. Moreover V_ω is tangent at $(0,0)$ to the linear space l_ω. If the terms $P_1(x,y,\omega)$ and $P_2(x,y,\omega)$ in (10.20) are of class C^r in (x,y) and if all derivatives of order up to and including r are jointly continuous in (x,y,ω), then V_ω is of class C^r. The manifolds V_ω vary continuously in ω. They are C^r-smooth in parameters $\lambda \in \mathbf{R}^p$ if A, B, Q, R, and P_1 and P_2 are C^r-smooth in λ. If Ω is a manifold, if A, B, Q, R, P_1, P_2 are C^r-smooth in ω, and if

$$(r+1)N < \frac{\alpha}{2},$$

then the V_ω vary C^{r-1}-smoothly in ω. Here α is the dichotomy constant mentioned above, and N is the "largest Lyapunov exponent" of the flow on Ω. More precisely, let d be a metric on Ω generating its topology, then it is required that N satisfy

$$d(T_t(\omega_1), T_t(\omega_2)) \leq d(\omega_1, \omega_2)e^{N|t|}$$

for all $t \in \mathbf{R}$ and $\omega_1, \omega_2 \in \Omega$.

It is now clear how to obtain y_0 and the corresponding exponentially decaying solution $\begin{pmatrix} x(t) \\ y(t) \end{pmatrix}$ of (10.20). By tangency of V_ω to l_ω, V_ω is a graph over the x-space $\mathbf{R}^n$, and it is the case that all these graphs are defined in some ball $B_{\rho_1}(0) \subset \mathbf{R}^n$ where ρ_1 does not depend on ω. Let $x_0 \in B_{\rho_1}(0)$; then there exists a unique $y_0 = y_0(\omega) \in \mathbf{R}^n$ such that $\begin{pmatrix} x_0 \\ y_0 \end{pmatrix} \in V_\omega$. Using (10.18) we get a feedback

$$u = R^{-1}(\omega)B^*(\omega)y_0(\omega) + \mathcal{G}(x,y,\omega)$$

such that $u(t)$ tends to zero exponentially as $t \to \infty$ and has smoothness properties with respect to parameters λ and with respect to ω which correspond to those listed in the preceding paragraph. The component $x(t)$ of the solution $\begin{pmatrix} x(t) \\ y(t) \end{pmatrix}$ of (10.20) is an exponentially decreasing solution of the control process (10.12). Clearly u satisfies the Pontryagin condition, and the functional I is finite upon substitution of $x(t)$ and $u(t)$.

We summarize these results in the following.

10.2 Theorem. *Consider the functional*

$$I_\omega(x,u) = \int_0^\infty L(x(t), u(t), T_t(\omega))dt$$

with

$$L(x,u,\omega) = \frac{1}{2}[\langle x, Q(\omega)x\rangle + \langle u, R(\omega)u\rangle] + \hat{L}(x,u,\omega),$$

where $Q, R, \hat{L}$ *satisfy the conditions enunciated above (see* (10.10) *and* (10.11)*). Then there exists a jointly continuous function*

$$(x,\omega) \to \hat{F}(x,\omega) : W \times \Omega \to M(m,n)$$

(where $W \subset \mathbf{R}^n$ *is a neighbourhood of the origin) and a number* $\delta > 0$ *with the following properties.*

(1) *Let* $x(t)$ *be the solution of* (10.12) *with* $x(0) = x_0$ *and* $u = \hat{F}(x, T_t(\omega))$*, then* $|x(t)| \to 0$ *exponentially fast as* $t \to \infty$*. Moreover, the pair* $\big(x(t), u(t)\big)$ *where* $u(t) = \hat{F}(x(t), T_t(\omega))$ *renders finite the functional* I_ω*. The pair* (x,u) *satisfies the Pontryagin necessary condition:*

$$\frac{\partial H}{\partial u}(x(t), y(t), u(t), T_t(\omega)) = 0,$$

where H *is given in* (10.14) *and* $\begin{pmatrix} x(t) \\ y(t) \end{pmatrix}$ *is a certain solution of the Hamiltonian system defined by* H *(see* (10.16a-b)*). More precisely,* $\begin{pmatrix} x(t) \\ y(t) \end{pmatrix}$ *is the solution of* (10.16a-b) *such that* $\begin{pmatrix} x(0) \\ y(0) \end{pmatrix} = \begin{pmatrix} x_0 \\ y_0(x_0,\omega) \end{pmatrix}$*, where* $y_0(x_0,\omega)$ *is obtained using the fact that the manifold* V_ω *is a graph; (see above). The function* $(x_0,\omega) \to y_0(x_0,\omega)$ *is jointly continuous on* $W \times \Omega$*.*

(2) *If the terms* P_1, P_2 *of* (10.20) *are of class* C^r *in* (x,y)*, then* $x_0 \to y_0(x_0,\omega)$ *and* $x \to \hat{F}(x,\omega)$ *are* C^r*-smooth in* W*.*

(3) *Moreover, if* A, B, Q, R, P_1 *and* P_2 *are of class* C^r *in* ω*, and if* $(r+1)N < \frac{\alpha}{2}$*, where* N *is the largest Lyapunov exponent of the flow on* Ω*, then* y_0 *and* $\hat{F}$ *are of class* C^{r-1} *in* ω*.*

Proof. The proof consists in defining $\hat{F}(x,\omega) = F(x, y_0(x,\omega), \omega)$ and following the development given above. □

References

[AM] B. Anderson and J. Moore: Detectability and stabilizability of time-varying discrete time linear systems, SIAM Journal of Control and Optimization 19, no.1 (1981), 20-32.

[Ar] V. Arnold: On a characteristic class entering in a quantum condition, Functional Analysis and its Applications 1 (1969), 1-14.

[At] F. Atkinson: Discrete and Continuous Boundary Value Problems, Academic Press, New york, London (1964).

[As] Z. Artstein: Uniform controllability via limiting systems, Applied Mathematics and Optimization 9 (1982), 111-131.

[B] A. Bacciotti: Local Stabilizability of Nonlinear Control Systems, World Scientific, Singapore, 1992.

[Bo1] P. Bougerol: Kalman filtering with random coefficients and contraction, SIAM Journal of Control and Optimization 31, no.4 (1993), 942- 959.

[Bo2] P. Bougerol: Filtre de Kalman Bucy at exposants de Lyapunov, preprint (1993).

[BS1] B. Barmish and W. Schmittendorf: Null controllability of linear systems with constrained controls, SIAM Journal of Control and Optimization 18, no.4 (1980), 327-345.

[BS2] B. Barmish and W. Schmittendorf: A necessary and sufficient condition for local constrained controllability of linear systems, IEEE Transactions in Automatic Control 25, no.1 (1980), 97-100.

[BT] I. Blinov and E. Tonkov: Global controllability of conditionally periodic linear systems, Trans. Mat. Zametki 32 (1982).

[Ch] V. Cheng: A direct way to stabilize continuous time and discrete time linear time varying systems, Transactions of Automatic Control AC-24 (1979), 641-643.

[CJ] F. Colonius and R. Johnson, Local and global null controllability of time varying linear control systems, submitted (1996).

[Cn] C. Conley: Isolated invariant sets and the Morse index, CBMS Regional Conference Series in Mathematics 38 (1978).

[Cp] A. Coppel: Dichotomies in Stability theory, lecture Notes in Mathematics 629, Springer-Verlag, Heidelberg (1978).

[CS] C. Chicone and R. Swanson: A generalized Poincare stability criterion, Procedings of the AMS 81, no.3 (1981), 495-500.

[CY] S.-N. Chow and Y.-F. Yi: Center manifold and stability for skew-product flows, CDSNS preprint 153, Georgia Tech. University (1993).

[DI] G. DaPrato and A. Ichikawa: Quadratic control for linear time varying systems, SIAM Journal of Control and Optimization 28, no.2 (1990), 359-381.

[E] R. Ellis: Lectures on Topological Dynamics, Benjamin, New York (1969).

[EJ] R. Ellis and R. Johnson: Topological dynamics and linear differential systems, Journal of Differential Equations 44 (1982), 21-39.

[H] R. Hermann: Cartanian Geometry, Nonlinear Waves and Control Theory, Part A, Interdisciplinary Mathematics Series, Vol. XX, Math Sci Press (1980).

[IMK] M. Ikeda, H. Maeda and S. Kodama: Stabilization of linear systems, SIAM Journal of Control 10 (1972), 716-729.

[J1] R. Johnson: Minimal functions with unbounded integral, Israel Journal of Mathematics 31 (1978), 133-141.
[J2] R. Johnson: Ergodic theory and linear differential equations, Journal of differential Equations 28 (1978), 23-34.
[J3] R. Johnson: Analyticity of spectral subbundles, Journal of differential Equations 35 (1980), 366-387.
[J4] R. Johnson: m-functions and Floquet exponents for linear differential systems, Annali di Mat. Pura ed Appl. 147 (1987), 211-248.
[JN1] R. Johnson and M. Nerurkar: On null controllability of linear systems with recurrent coefficients and constrained controls, Journal of Dynamics and Differential Equations 4 (1992), 259-273.
[JN2] R. Johnson and M. Nerurkar: Feedback control for linear chaotic systems, Proceedings of the 2nd IFAC Workshop on System Structure and Control, Prague (1992) 272-273.
[JN3] R. Johnson and M. Nerurkar: Exponential dichotomy and rotation number for linear Hamiltonian systems, Journal of Differential Equations 108 (1994), 201-216.
[JN4] R. Johnson and M. Nerurkar: Null controllability of linear systems with positive Lyapunov exponents, Differential Equations, Dynamical Systems and Control Science, edited by K. Elworthy and W. Everitt, Marcel Dekker Publishers (1994), 605-621.
[JN5] R. Johnson and M. Nerurkar: Stabilization and random linear regulator problem for random linear control processes, Journal of Mathematical Analysis and Applications 197 (1996), 608-629.
[JPS] R. Johnson, K. Palmer and G. Sell: Ergodic properties of linear dynamical systems, SIAM Journal of Mathematical Analysis 18, no.1 (1987), 1-33.
[K] R. Kalman: Contributions to the theory of optimal control, Bull. Soc. Math. Mex. 5 (1960),102-119.
[LM] E. Lee and L. Marcus: Foundations of Optimal Control Theory, John Wiley and Sons, New York (1967).
[M1] V. Millionschikov: Proof of the existence $\cdots$ almost periodic coefficients, Differential Equations 4(1968), 203-205.
[M2] V. Millionschikov: Typicality of almost reducible systems with almost periodic coefficients, Differential Equations 14 (1978), 448-450.
[MS] R. Miller and G. Sell: Volterra integral equations and topological dynamics, AMS Memoirs 102, Providence R.I. (1970).
[MSc] J. Massera and J. Schaeffer: Linear Differential Equations and Function Spaces, Academic Press, New York, London (1966).
[N1] M. Nerurkar: An example of almost universally observable vector field on $\mathbf{P}^1(\mathbf{R})$, Systems and Control Letters 14 (1990),
[N2] M. Nerurkar: Observability and topological dynamics, Journal of Dynamics and Differential Equations 3 , no.2 (1991), 273-287.
[N3] M. Nerurkar: Controllability and the nature of quasi- frequencies, System and Control Letters 16 (1991), 195-198.
[NS] V. Nemytskii and V. Stepanov: Qualitative Theory of Ordinary Differential Equations, Princeton University Press (1960).

[P1] K. Palmer: On reducibility of almost periodic systems of linear differential equations, Journal of Differential Equations 35 (1980), 374-380.
[P2] K. Palmer: Two linear systems criteria for exponential dichotomy, Annali Mat. a Pura ed Appl. 124 (1980), 199-216.
[P3] K. Palmer: Exponential dichotomoies and transversal homoclinic points, Journal of Differential Equations 55 (1984), 225-256.
[P4] K. Palmer: Transversal heteroclinic points and Cherry's example of a non-integrable Hamiltonian system, Journal of Differential Equations 65 (1986), 321-360.
[Pa] L. Pandolfi: Linear control systems: Controllability with constrained control, Journal of Optimization Theory and Applications 19 (1976).
[PBGM] L. Pontryagin, V. Baltyanskii, R. Gamkrelidze and E. Mishchenko: The Mathematical Theory of Optimal Processes, John Wiley and Sons (Interscience), New York (1962).
[Sc] I. Ya. Schneiberg: Zeros of integrals along trajectories of ergodic systems, Trans. Funk. Anal. Prilozh. 19 (1985), 486-490.
[Se] G. Sell: Topological Dynamics and Ordinary Differential Equations, Van Nostrand-Reinhold, London (1971).
[Sg] J. Selgrade: Isolated invariant sets for flows on vector bundles, Transaction of AMS 203 (1975), 359-390.
[SS] R. Sacker and G. Sell: A spectral theory of linear differential systems, Journal of Differential Equations 27 (1978), 320-358.
[T] E. Tonkov: Dynamic system of translations and questions of uniform controllability of linear systems, Soviet Math. Doklady 23 (1981).
[Y] Y.-F. Yi: A generalized integral manifold theorem, Journal of Differential Equations 102 (1993), 153-187.

Departimento di Sistemi e Informatica, Universita di Firenze, Firenze 50139, Italy.
Rutgers University, Department of Mathematics, Camden NJ 08102.

Editorial Information

To be published in the *Memoirs*, a paper must be correct, new, nontrivial, and significant. Further, it must be well written and of interest to a substantial number of mathematicians. Piecemeal results, such as an inconclusive step toward an unproved major theorem or a minor variation on a known result, are in general not acceptable for publication. *Transactions* Editors shall solicit and encourage publication of worthy papers. Papers appearing in *Memoirs* are generally longer than those appearing in *Transactions* with which it shares an editorial committee.

As of July 31, 1998, the backlog for this journal was approximately 6 volumes. This estimate is the result of dividing the number of manuscripts for this journal in the Providence office that have not yet gone to the printer on the above date by the average number of monographs per volume over the previous twelve months, reduced by the number of issues published in four months (the time necessary for preparing an issue for the printer). (There are 6 volumes per year, each containing at least 4 numbers.)

A Copyright Transfer Agreement is required before a paper will be published in this journal. By submitting a paper to this journal, authors certify that the manuscript has not been submitted to nor is it under consideration for publication by another journal, conference proceedings, or similar publication.

Information for Authors and Editors

Memoirs are printed by photo-offset from camera copy fully prepared by the author. This means that the finished book will look exactly like the copy submitted.

The paper must contain a *descriptive title* and an *abstract* that summarizes the article in language suitable for workers in the general field (algebra, analysis, etc.). The *descriptive title* should be short, but informative; useless or vague phrases such as "some remarks about" or "concerning" should be avoided. The *abstract* should be at least one complete sentence, and at most 300 words. Included with the footnotes to the paper, there should be the 1991 *Mathematics Subject Classification* representing the primary and secondary subjects of the article. This may be followed by a list of *key words and phrases* describing the subject matter of the article and taken from it. A list of the numbers may be found in the annual index of *Mathematical Reviews*, published with the December issue starting in 1990, as well as from the electronic service e-MATH [**telnet e-MATH.ams.org** (or **telnet 130.44.1.100**). Login and password are **e-math**]. For journal abbreviations used in bibliographies, see the list of serials in the latest *Mathematical Reviews* annual index. When the manuscript is submitted, authors should supply the editor with electronic addresses if available. These will be printed after the postal address at the end of each article.

Electronically prepared papers. The AMS encourages submission of electronically prepared papers in $\mathcal{A}\mathcal{M}\mathcal{S}$-TeX or $\mathcal{A}\mathcal{M}\mathcal{S}$-LaTeX. The Society has prepared author packages for each AMS publication. Author packages include instructions for preparing electronic papers, the *AMS Author Handbook*, samples, and a style file that generates the particular design specifications of that publication series for both $\mathcal{A}\mathcal{M}\mathcal{S}$-TeX and $\mathcal{A}\mathcal{M}\mathcal{S}$-LaTeX.

Authors with FTP access may retrieve an author package from the Society's Internet node `e-MATH.ams.org` (130.44.1.100). For those without FTP access, the author package can be obtained free of charge by sending e-mail to `pub@ams.org` (Internet) or from the Publication Division, American Mathematical Society, P.O. Box 6248, Providence, RI 02940-6248. When requesting an author package, please specify AMS-TeX or AMS-LaTeX, Macintosh or IBM (3.5) format, and the publication in which your paper will appear. Please be sure to include your complete mailing address.

Submission of electronic files. At the time of submission, the source file(s) should be sent to the Providence office (this includes any TeX source file, any graphics files, and the DVI or PostScript file).

Before sending the source file, be sure you have proofread your paper carefully. The files you send must be the EXACT files used to generate the proof copy that was accepted for publication. For all publications, authors are required to send a printed copy of their paper, which exactly matches the copy approved for publication, along with any graphics that will appear in the paper.

TeX files may be submitted by email, FTP, or on diskette. The DVI file(s) and PostScript files should be submitted only by FTP or on diskette unless they are encoded properly to submit through e-mail. (DVI files are binary and PostScript files tend to be very large.)

Files sent by electronic mail should be addressed to the Internet address `pub-submit@ams.org`. The subject line of the message should include the publication code to identify it as a Memoir. TeX source files, DVI files, and PostScript files can be transferred over the Internet by FTP to the Internet node `e-math.ams.org` (130.44.1.100).

Electronic graphics. Figures may be submitted to the AMS in an electronic format. The AMS recommends that graphics created electronically be saved in Encapsulated PostScript (EPS) format. This includes graphics originated via a graphics application as well as scanned photographs or other computer-generated images.

If the graphics package used does not support EPS output, the graphics file should be saved in one of the standard graphics formats—such as TIFF, PICT, GIF, etc.—rather than in an application-dependent format. Graphics files submitted in an application-dependent format are not likely to be used. No matter what method was used to produce the graphic, it is necessary to provide a paper copy to the AMS.

Authors using graphics packages for the creation of electronic art should also avoid the use of any lines thinner than 0.5 points in width. Many graphics packages allow the user to specify a "hairline" for a very thin line. Hairlines often look acceptable when proofed on a typical laser printer. However, when produced on a high-resolution laser imagesetter, hairlines become nearly invisible and will be lost entirely in the final printing process.

Screens should be set to values between 15% and 85%. Screens which fall outside of this range are too light or too dark to print correctly.

Any inquiries concerning a paper that has been accepted for publication should be sent directly to the Editorial Department, American Mathematical Society, P. O. Box 6248, Providence, RI 02940-6248.

Editors

This journal is designed particularly for long research papers (and groups of cognate papers) in pure and applied mathematics. Papers intended for publication in the *Memoirs* should be addressed to one of the following editors:

Selected Titles in This Series

(Continued from the front of this publication)

617 **Robin Hartshorne,** Families of curves in $\mathbb{P}^3$ and Zeuthen's problem, 1997

616 **Serguei G. Bobkov and Christian Houdré,** Some connections between isoperimetric and Sobolev-type inequalities, 1997

615 **Michael A. Dritschel and Hugo J. Woerdeman,** Model theory and linear extreme points in the numerical radius unit ball, 1997

614 **Richard Warren,** The structure of k-CS-transitive cycle-free partial orders, 1997

613 **D. L. Flannery,** The finite irreducible linear 2-groups of degree 4, 1997

612 **Joan Porti,** Torsion de Reidemeister pour les variétés hyperboliques, 1997

611 **D. Ginzburg, I. Piatetski-Shapiro, and S. Rallis,** L functions for the orthogonal group, 1997

610 **Mark Hovey, John H. Palmieri, and Neil P. Strickland,** Axiomatic stable homotopy theory, 1997

609 **Liviu I. Nicolaescu,** Generalized symplectic geometries and the index of families of elliptic problems, 1997

608 **Christina Q. He and Michel L. Lapidus,** Generalized Minkowski content, spectrum of fractal drums, fractal strings, and the Riemann zeta-functions, 1997

607 **Adele Zucchi,** Operators of class C_0 with spectra in multiply connected regions, 1997

606 **Moshé Flato, Jacques C. H. Simon, and Erik Taflin,** Asymptotic completeness, global existence and the infrared problem for the Maxwell-Dirac equations, 1997

605 **Liangqing Li,** Classification of simple C^*-algebras: Inductive limits of matrix algebras over trees, 1997

604 **Hajnal Andréka, Steven Givant, and István Németi,** Decision problems for equational theories of relation algebras, 1997

603 **Bruce N. Allison, Saeid Azam, Stephen Berman, Yun Gao, and Arturo Pianzola,** Extended affine Lie algebras and their root systems, 1997

602 **Igor Fulman,** Crossed products of von Neumann algebras by equivalence relations and their subalgebras, 1997

601 **Jack E. Graver and Mark E. Watkins,** Locally finite, planar, edge-transitive graphs, 1997

600 **Ambar Sengupta,** Gauge theory on compact surfaces, 1997

599 **Tai-Ping Liu and Yanni Zeng,** Large time behavior of solutions for general quasilinear hyperbolic-parabolic systems of conservation laws, 1997

598 **Valentina Barucci, David E. Dobbs, and Marco Fontana,** Maximality properties in numerical semigroups and applications to one-dimensional analytically irreducible local domains, 1997

597 **Ragnar-Olaf Buchweitz and John J. Millson,** CR-geometry and deformations of isolated singularities, 1997

596 **Paul S. Bourdon and Joel H. Shapiro,** Cyclic phenomena for composition operators, 1997

595 **Eldar Straume,** Compact connected Lie transformation groups on spheres with low cohomogeneity, II, 1997

594 **Solomon Friedberg and Hervé Jacquet,** The fundamental lemma for the Shalika subgroup of $GL(4)$, 1996

593 **Ajit Iqbal Singh,** Completely positive hypergroup actions, 1996

592 **P. Kirk and E. Klassen,** Analytic deformations of the spectrum of a family of Dirac operators on an odd-dimensional manifold with boundary, 1996

(See the AMS catalog for earlier titles)